RELIABILITY FUNDAMENTALS

FUNDAMENTAL STUDIES IN ENGINEERING

Vol. 1 Coelho/Physics of Dielectrics for the Engineer
Vol. 2 Szücs/Similitude and Modelling
Vol. 3 Scharf, Lisieski/Amplitude Distribution Spectrometers
Vol. 4 Mohilla, Ferenz/Chemical Process Dynamics
Vol. 5 Baker et al./Explosion Hazards and Evaluation
Vol. 6 Sobczyk/Stochastic Wave Propagation
Vol. 7 König/Shakedown of Elastic-Plastic Structures
Vol. 8 Čačko, Bílý, Bukoveczky/Random Processes
Vol. 9 Kubáček/Foundations of Estimation Theory
Vol. 10 Cătuneanu, Mihalache/Reliability Fundamentals

FUNDAMENTAL STUDIES IN ENGINEERING 10

RELIABILITY FUNDAMENTALS

Vasile M. CĂTUNEANU
Polytechnic Institute, Bucharest, Romania

Adrian N. MIHALACHE
Polytechnic Institute, Bucharest, Romania

ELSEVIER
Amsterdam — Oxford — New York — Tokyo 1989

Revised translation of the Romanian book
BAZELE TEORETICE ALE FIABILITĂŢII
by V. Cătuneanu and A. Mihalache
Editura Academiei Republicii Socialiste România, Bucharest, 1983

Translated from the Romanian by Adrian Mihalache

Published in co-edition by
EDITURA ACADEMIEI, Bucharest, and
ELSEVIER SCIENCE PUBLISHERS, Amsterdam

Distribution of this book is being handled by the following publishers:

for the U.S.A. and Canada

ELSEVIER SCIENCE PUBLISHING COMPANY, INC.
655 Avenue of the Americas
New York, NY 10010, U.S.A.

for the East European Countries, China, Northern Korea, Cuba, Vietnam and Mongolia

EDITURA ACADEMIEI REPUBLICII SOCIALISTE ROMÂNIA
Calea Victoriei 125
R-79717 Bucharest, Romania

for all remaining areas

ELSEVIER SCIENCE PUBLISHERS
Sara Burgerhartstraat 25
P.O. Box 211, 1000 AE Amsterdam, The Netherlands

Library of Congress Cataloging-in-Publication Data

Cătuneanu, Vasile M.
[Bazele teoretice ale fiabilităţii. English]
Reliability fundamentals / Vasile M. Cătuneanu, Adrian N. Mihalache.
p. cm.——(Fundamental studies in engineering; 10)
Rev. and updated translation of: Bazele teoretice ale fiabilităţii. 1983.
Translated from Romanian by Adrian Mihalache.
Includes bibliographies and index.
1. Reliability (Engineering) I. Mihalache, Adrian. II. Title. III. Series.
TA169.C3813 1989
620′.00452——dc 19 88—38601
CIP

ISBN 0-444-98879-3 (Vol. 10)
ISBN 0-444-41756-7 (Series)
ISBN 973-27-0052-1 (Romania)

PRINTED IN ROMANIA

Contents

Abbreviations

ARP	Age Replacement Policy
BRP	Block Replacement Policy
cdf	cumulative distribution function
CFR	Constant Failure Rate
CRP	Continuous Replacement Policy
cv	coefficient of variation
DFR	Decreasing Failure Rate
DFRA	Decreasing Failure Rate Average
DMRL	Decreasing Mean Residual Life
DRP	Delayed Replacement Policy
ERP	Eventual Replacement Policy
FRP	Failure Replacement Policy
iff	if and only if
IFR	Increasing Failure Rate
IFRA	Increasing Failure Rate Average
iid	independent and identically distributed
IMRL	Increasing Mean Residual Life
LCC	Life-Cycle Costing
ML	Maximum Likelihood
MLE	Maximum Likelihood Estimation
MRL	Mean Residual Life
MSE	(and mse) Mean Square Error
MTBF	Mean Time Between Failures
MTTF	Mean Time To Failure
MTTFF	Mean Time To First Failure
MTTR	Mean Time To Repair
NBU	New Better than Used
NBUE	New Better than Used in Expectation
NWU	New Worse than Used
NWUE	New Worse than Used in Expectation
O&S	Operations and Support
pdf	probability density function
R&D	Research and Development
RDT&E	Research Development Testing and Evaluation
s–d–p	sum of disjoint products
s–o–p	sum of products
UMVUE	Uniformly Minimum Variance Unbiased Estimator

Chapter 1

Reliability in the framework of general system theory

Reliability is an inter-disciplinary concept for treating the degradation laws of physical elements and systems. Although attention has long been focused on degradation phenomena, reliability, as a theory and a distinct field of research, was founded only some thirty years ago. A quick glimpse into the main ideas of reliability theory may be obtained by considering the circumstances of its birth.

The technological developments which accompanied the crest of the second wave of human civilization (in Toffler's terms [1]) led to the building of large technical systems, of great complexity and therefore prone to frequent failures, which often cause considerable damage. It is sufficient to consider the systems for space research, modern weapons and nuclear power plants to appreciate the necessity of taking all security measures against any possible undesirable event.

Technical means must be employed to ensure proper behaviour of the systems under specified stresses and over specified time intervals.

The numerous examples of serious incidents caused by the failure of technical systems of great complexity [2] show that apparently minor causes can have major effects, which are to a large extent unpredictable.

Apart from the complexity of the system, the acceleration of the degradation process is a natural consequence of the mass production related to the general level of economic development. The adverse effects of mass production on quality were noticed long ago, and substantial efforts have been made to overcome them, but until the late 1940s there was no clear distinction between quality assurance at a certain moment and the conservation of system performance over a determined time span. It was tacitly assumed that, if a system behaves well now, it will do so indefinitely. This supposition was supported by the evidence of the technological tradition. The methodology of design based on stress-strength considerations guaranteed, in principle, conservation of the system performance over a long enough time interval, given normal maintenance. But this methodology may be employed only in those technical areas where a vast experience of material properties and technological know-how has been accumulated. The high rate of change of modern electronic technology prevents accumulation of the experience which is essential in order to develop design methods to assure the conservation of system performance in a deterministic way.

The development of modern technologies enlarged the scope of the notion of quality, to include aspects concerning the time varying performances of systems. The general property of a system, to conserve its performance in time, became the specific notion, that is reliability. The concept of reliability, developed on the basis of electronic technology, was readily generalized to describe any system, be it technical, biological or even social. This book is concerned with technological systems, but many of its ideas can be applied successfully outside the purely technical field.

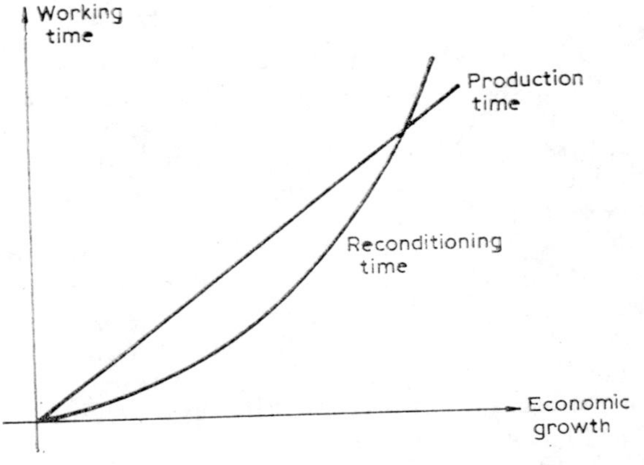

Fig. 1.1. — Working time for production and reconditioning as functions of economic growth.

Reliability, as a general system property, provides a particular understanding of the future of material production [3, 4]. If a constant level of reliability is assumed, the working time required for production increases linearly with the level of economic development, while the time required for the weeding-out and correction of faulty products increases exponentially (Fig. 1.1). Thus a critical level of development could be reached when the time for reconditioning of goods would equal the time for their original production. This means that half of the active population would be busy repairing goods produced by the other half. A continuous improvement of reliability must accompany economic development in order to avoid such a critical situation.

Improvement of system reliability is unlikely if the present trend towards mass production and greater complexity persists. One might imagine that the prospective development of simpler, adaptive systems, together with new conceptions of the production process typical of the third wave of civilization [1], would have a beneficial influence on reliability. However, only conscious human action may counteract the natural evolution of systems towards degradation.

The above considerations explain the present interest in reliability as a practical matter, rather than a field of theoretical research.

The reasons for a theoretical approach to reliability are connected with the complexity of the system and with the features of modern mass production.

However high the investment on reliability of a complex system may be, perfect reliability cannot be achieved. One must at least know the actual level of reliability, in order to specify the length of an operation which may be accomplished safely, the renewal strategy which must be employed, the logistics, the safety factors, and so on.

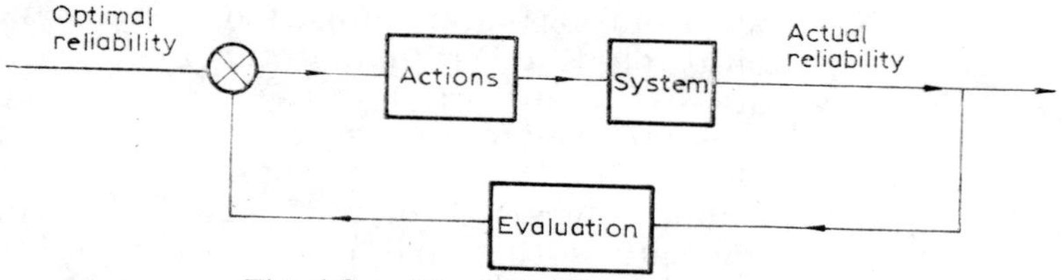

Fig. 1.2. — The reliability control loop.

Only a theoretical approach can provide a mathematical model to express the reliability performance quantitatively and furnish a methodology for deriving the parameters of the model from experimental data and knowledge of the system structure.

Considering now the features of mass production and taking into account the rapid changes determined by developing technologies, one can see that instead of striving to attain a very high reliability, it is more desirable to maintain a firm control on the reliability level, in order to set it to an optimal value. Such an approach is well represented by a feedback loop as shown in Fig. 1.2.

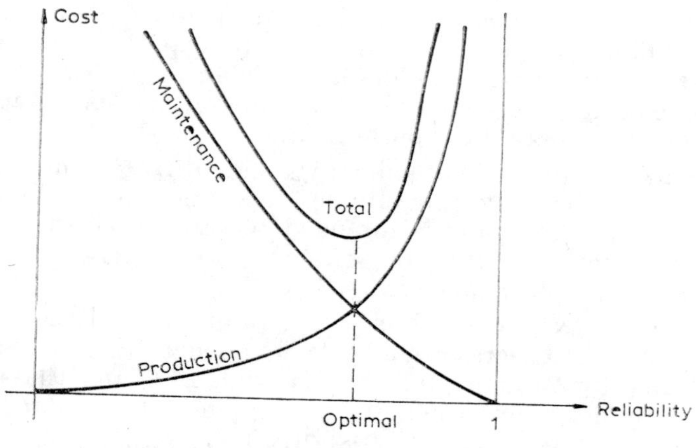

Fig. 1.3. — The optimal reliability level.

In order to keep the reliability under control, a methodology to evaluate the actual reliability level, and a cost model, are both required.

The optimal level of reliability may be established from an economic point of view. If the same manager is responsible for the production and the maintenance of a system, the optimal level may be derived by minimizing the sum of the production and maintenance costs (Fig. 1.3).

Another approach to the specification of a reliability goal is the methodology of life-cycle costing (LCC) [5, 6]. Life-cycle costing refers to an analysis technique which encompasses the total cost of a system over a specified period in its lifetime. This cost is the sum of (a) research, development, testing and evaluation (RDT&E) costs; (b) acquisition costs; (c) operations and support (O&S) costs. Reliability plays a very important role in all three areas. The proportion of LCC typically falling in each area is shown in Fig. 1.4. The most important observation is that operations and support costs comprise over half of LCC. Also, research, development, testing and evaluation costs comprise a relatively small part of LCC, yet have a great impact on the total cost. Programmes to improve reliability are classified under RDT&E. The proportion of total RDT&E costs normally consumed in a reliability investment programme ranges from 2% to 8% with a mean of 5%. The acquisition cost is influenced by reliability changes in two ways: (a) the cost of the labour spent in rectifying failures of systems occurring during the manufacturing process; (b) the cost of high-reliability components employed in the manufacture of the systems. Increasing the reliability of a system causes the mean number of system failures during manufacturing to fall commensurately. The influence on the unit cost of the use of high reliability components is to increase the cost of materials. Finally, the operating and support costs, dealing with maintenance man-hours and spare parts consumption, are sensitive to reliability changes. This is plausible to the extent that there is a one-to-one correlation between the number of system failures and the number of system repair jobs performed.

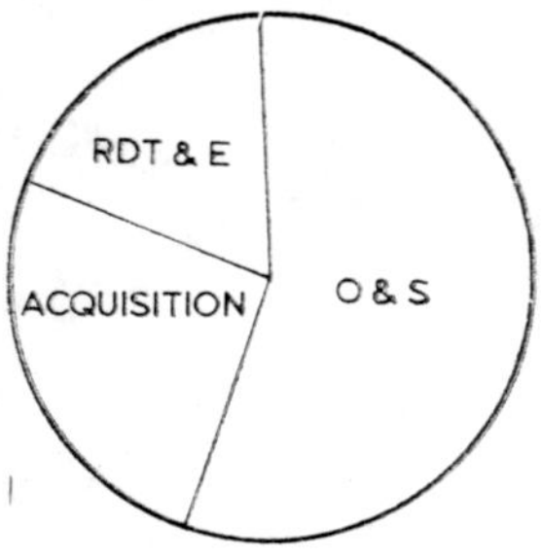

Fig. 1.4. — LCC breakdown into major cost areas.

Seger [6] develops a methodology to identify the optimal level of investment in a reliability programme such that the system life-cycle cost is minimized. The model is intended to substantiate the funding of reliability programmes during the system development phase.

The reliability investment is expressed as a fraction of the research and development (R&D) costs. For each reliability level the LCC is computed. The plot of LCC versus reliability investment (Fig. 1.5) shows that a low reliability investment gives a very high LCC and that LCC is insensitive to reliability investment in the region of minimum LCC.

The assurance of a desired reliability level implies the existence of a theory which permits the evaluation and control of reliability, based on mathematical models and on an adequate methodology. Such a theory, despite its advantages, is likely to impoverish the general meaning of reliability, circumventing it by using an unambiguous but restrictive mathematical definition. A necessary distinction must be made between reliability as a general property and the associated mathematical concept which only partially expresses it.

Together with a methodology for the practical activity of reliability assurance, the general theory of reliability engenders an essentially diachronic point of view, which seems particularly fertile now, at the end of

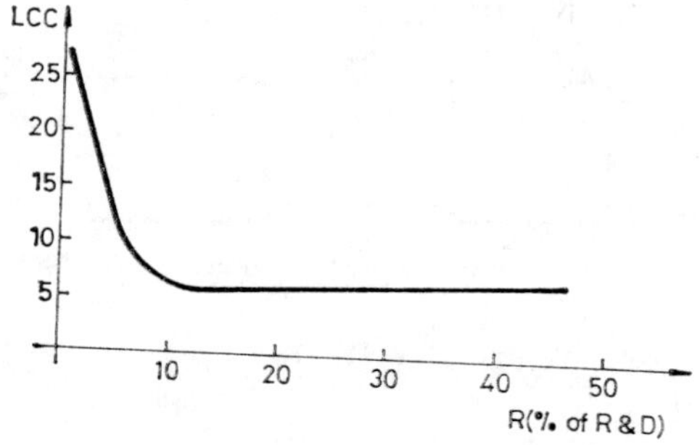

Fig. 1.5. — LCC vs reliability investment.

a period dominated by the synchronic, structuralist conception. The influence of this conception is responsible for the fact that the reliability model is expressed in the structuralist language of the general system theory.

According to this theory, a physical system is described by a set of variables (measurable properties) which represent its only connections to the outside world. The causality condition leads to a classification of the variables as input variables (causes) and output variables (effects). An abstract object is defined as a set of ordered pairs of time functions :

$$A = \{U_{[t_0, t_1]}, \ Y_{[t_0, t_1]}\} \tag{1.1}$$

which satisfy the concatenarity condition [7]. The vector $U = (u_1, u_2, \ldots, u_n)$ is the input vector and the vector $Y = (y_1, y_2, \ldots, y_p)$ is the output vector. The abstract object is thus a set of input-output pairs. A real system may be viewed as one abstract object at the global level, when only input and output variables are considered. The same system may be viewed at the structural level as an inter-connection of abstract objects. Now the elements of the system and their inter-relationships will also be considered. In the following we shall use the notion of an abstract system as the model of a real system, irrespective of the fact that it includes one or more abstract objects.

A system (whose structure may be ignored) is mathematically described at the global level by the functional dependence between the output and input vectors :

$$Y_{[t_0, t_1]} = A(U_{[t_0, t_1]}) \tag{1.2}$$

where A is an operator.

Some knowledge, even partial knowledge, of the structure of the system reveals a set of internal variables which mediate the $U \to Y$ causality. This set of variables is the state vector $X = (x_1, x_2 \dots x_n)$ of the system. A suggestive description of the relation between the input, output and state vectors is given in Fig. 1.6. The system is thus divided into two cascaded subsystems, the state variables being outputs from the first subsystem and inputs to the second.

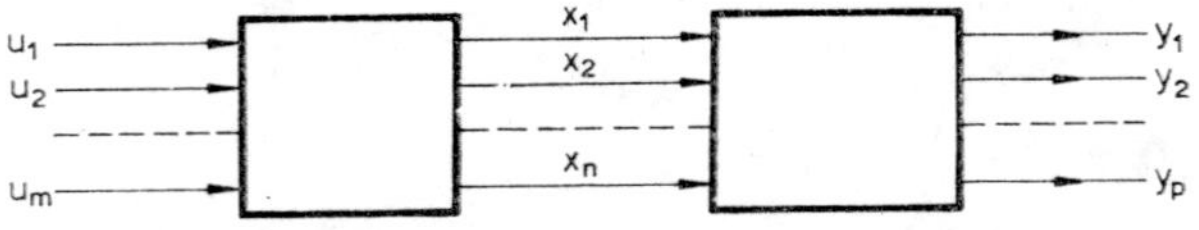

Fig. 1.6. — Input, output and state variables.

According to [6], almost any system may be described by the canonical state equations :

$$\dot{X} = f(t, X, U);$$
$$Y = g(t, X). \tag{1.3}$$

From these equations it may be seen that the state vector completely determines the output vector, so that the input-output dependence appears only through the state vector. The equations (1.3) are valid for a suitable choice of the state variables such that the state-output subsystem would be trivial [7].

The general model provided by the system theory approach may describe a system from different points of view, depending on the physical interpretations of the input, output and state variables. Let us consider an electrical network which is composed of resistors, capacitors, inductors, and independent and controlled sources. If we identify the input variables by the independent sources and the output variables by the potentials of different nodes, the abstract system will describe the response of the real systems to signals applied, according to the global relation (1.2). If the currents in the inductors and the voltages on the capacitors are chosen as state variables, the network will be structurally described by the canonical equations (1.3).

To obtain a reliability model of the network, another interpretation of the variables is required. The goal of reliability analysis is to describe the evolution of the system performance under specified stress conditions. It is then natural to consider as input variables the stresses to which the system is subjected, whether they are essential to the functioning of the system, such as potentials, currents and dissipated power, or perturbing stresses like ambient temperature, relative humidity and so on. The output variables describe the performance of the system. In the case of the electrical network, such variables are the transit time, stability, bandwidth etc. As the dependence of the performance on the stresses is more often than not mediated by the component or subsystem parameters, these parameters must be chosen as state variables in a reliability model. It is convenient to consider as output and state variables the relative deviations of the performance and parameter values from their nominal values. In the case of the electrical network, the state vector includes the values of the inductances, capacitances and resistances or their fractional deviations.

A global analysis of the reliability leads to a relationship of the type (1.2) between stresses and performance, whilst a structural analysis leads to canonical equations of the type (1.3) which describe the influence of the stresses on the performance, mediated through the parameters of the components.

The above considerations show the applicability of the general system model to reliability studies. A specific feature of the reliability model is the stochastic interdependence between its variables. Even when the stresses are deterministic values, as under laboratory conditions, the evolution of the performance is not entirely predictable, so that the output vector should be considered as a p-dimensional stochastic process. Similarly, the evolution of the component parameters, which form the state vector, may be described using the theory of stochastic processes. Moreover, the irreversibility of the degradation process implies that the stochastic processes associated with the output and state variables are non-stationary.

A quantitative analysis of the model is very laborious, as it implies the simultaneous description of p output stochastic processes and n state stochastic processes. Nevertheless, such an analysis is performed for measurement devices [8], where the performance vector includes the errors over different sections of the measuring scale.

In most cases, analysis of the output and state stochastic processes is avoided by use of the concept of failure. A failure is more rigorously defined in reliability theory than in common language. In order to define it, a subdomain Da of the p-dimensional space is assumed, so that the system functions satisfactorily as long as the performance vector belongs to Da, and otherwise it fails. It is also assumed that initially (at $t = 0$) the performance vector belongs to the acceptable domain Da. The performance of the system changes stochastically, until the frontier of Da is crossed and system failure occurs. The failure process is shown in Fig. 1.7 for a one-dimensional output vector.

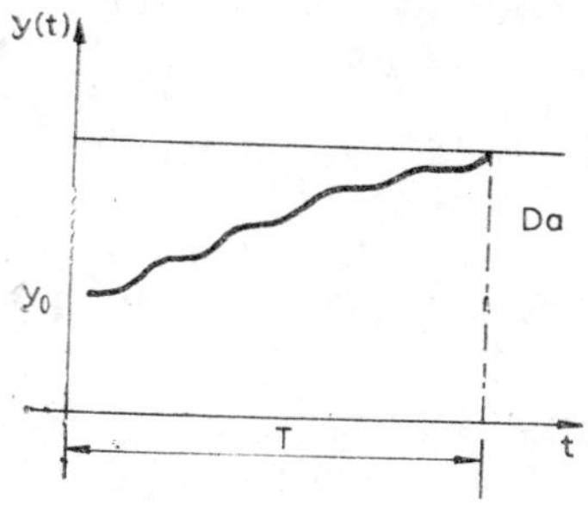

Fig. 1.7. — Failure process.

Note that the system reliability may be modelled either by the stochastic process $y(t)$ or by the time to failure of the system, which is a continuous random quantity. The global analysis may be effected using this random variable, irrespective of the number of stochastic processes involved in the output vector.

The concept of failure plays the same important role in the structural analysis, wherein the stochastic processes associated with the component parameters are replaced by their appropriate times to failure. Thus the goal of the structural analysis is to establish a relationship between random variables instead of stochastic processes.

The validity of the definition of failure rests on a correct specification of the acceptable domains associated with the system performance and with the component parameters. This implies a realistic definition of the system requirements and a correct tolerance analysis performed in the design phase.

All of the above methodological considerations neglect the representation of the input vector. Although the stresses are random quantities, the necessity of a simplified analysis leads to assuming constant, deterministic values of the stresses, as in laboratory conditions. The influence of the stresses on the component parameters and hence on the system performance may not be completely described in the framework of general theory. The physical and chemical processes of degradation, and the failure modes and mechanisms, must be studied as necessary stages in the analysis of specific systems. The experience accumulated up to the present time is concerned mainly with electronic devices, for which partially empirical laws describing the influence of stresses on the reliability have been derived. A general solution for the dynamic equation of the system reliability $\dot{X} = f(t, X, U)$, which models the dependence of the component parameters on the stresses applied, cannot yet be given.

This book aims to present the theory and practice of reliability to a sufficient extent that they may be applied to all kinds of systems. It contains the basic models and methods for reliability analysis at the global and at the structural level.

The next two chapters are concerned with global modelling, according to the general equation (1.2), of the evolution of system performance under constant stresses. The fourth chapter describes the relationship between reliability and stresses and presents a theoretical approach to accelerated life-testing. The fifth chapter deals with those models which account for external interventions intended to delay the natural degradation process and to improve the system reliability. The sixth chapter includes the models and methods employed in the structural analysis of reliability using the general equation $Y = g(t, X)$. The last chapter presents the general principles of reliability theory and places it in the context of contemporary scientific ideas.

References

1. TOFFLER, A., The Third Wave, Bantam Books, 1981.
2. NITU, V., Energetic Systems Reliability (in Romanian), Ed. Academiei, Bucharest, 1975.
3. CARTIANU, GH., CĂTUNEANU, V.M., Analyse zwecks Optimierung des Erneuerungsprozesses, Tagung 1975 Nürenberg Technische Zuverlässigkeit, Ausschuss technische Zuverlässigkeit VDI.

4. CARTIANU, GH., CĂTUNEANU, V. M., Contributions to the Study of the Relationship between Development, Consumption and Renewal. Researches in Electronic Technology and Reliability (in Romanian), Ed. didactică și pedagogică, Bucharest, 1979, pp. 12—15.
5. REGULINSKI, T. L., GUPTA, Y. P., Reliability cost estimation : managerial perspectives, IEEE Trans. Reliab., R-32, 3 (1983) 276—281.
6. SEGER, J. K., Reliability investment and life-cycle cost, IEEE Trans. Reliab., R-32, 3 (1983) 259—263.
7. ZADEH, L., POLAK, E., System Theory (in Romanian), Ed. Tehnică, Bucharest, 1974.
8. WIENER, UL., A New Mathematical Model and Algorithm for the Study of Intrinsic Reliability of Measuring Devices. Applied Technology (in Romanian), XXIII, 3(1976), 129 — 138.

Chapter 2

Global modelling of system reliability

2.1 Reliability measures

The mathematical description of the reliability of a system may be approached either from a global point of view, ignoring the structure of the systems, or from a structural level, taking into account the components of the system and their inter-relationships.

In both cases, the analysis is performed in a one-dimensional space considering the time interval elapsed since the initial moment of the system operation up to the moment of failure, when at least one of the system variables transgresses its prescribed limits. For non-repairable systems the analysis is restricted to the time interval to the first failure of the system, whilst for those that can be repaired, all time intervals between successive failures must be considered. For the latter category, similar reliability analyses are performed after each repair.

Let T be the time to failure of a system, whether it is the time to the first failure (in the case of non-repairable systems) or the time between (successive) failures (in the case of repairable systems). In both cases, T represents an interval during which the system is continuously working satisfactorily, given its good state at the beginning of the interval. The time to failure T is a random variable and its mathematical characteristics may be interpreted as components of a global reliability model. The model is a general one in the sense that it can describe both repairable and non-repairable systems, but for repairable systems, the model must be updated after each repair.

The fundamental mathematical characteristic of a continuous random variable T is the cumulative distribution function (cdf):

$$F(t) = \text{Prob}\,\{T < t\}. \tag{2.1}$$

From a reliability point of view, $F(t)$ represents the probability that the system fails in $(0, t)$. It is tacitly assumed that the system is good at $t = 0$, which is the initial working time or the time when a repair is completed.

The complement of $F(t)$ is the reliability function $R(t)$, which represents the probability that an operation of duration t, assigned to the system at $t = 0$ is successfully accomplished.

It is important to notice that $F(t)$ and $R(t)$ refer to the time interval $(0, t)$ and not to the end point t of the interval, as the notation seems to imply.

Let $(t, t + x)$ be a time interval included in the time to failure of a system. The probability of failure in $(t, t + x)$, by definition of cdf, is :

$$F(t, t + x) = \text{Prob}\,\{t \leqslant T < t + x\} = F(t + x) - F(t). \tag{2.2}$$

The reliability function in $(t, t + x)$ is then :

$$R(t, t + x) = 1 - F(t, t + x) = R(t + x) + F(t). \tag{2.3}$$

It seems, from expression (2.3), that an operation of duration x assigned at time t is successfully accomplished either if the system does not fail up to $t + x$ or if it fails before t. As this interpretation is contrary to reason, another measure of the reliability in the interval $(t, t + x)$ must be employed. Let $F(t, t + x/T \geqslant t) = F(x/t)$ be the probability of failure in $(t, t + x)$ given that the system has not failed up to t. By the definition of conditional probabilities, we get :

$$\begin{aligned} F(x/t) &= \frac{F(t, t + x)}{R(t)} = \frac{F(t + x) - F(t)}{R(t)}; \\ R(x/t) &= R(t, t + x/T \geqslant t) = 1 - F(x/t) = \frac{R(t + x)}{R(t)}. \end{aligned} \tag{2.4}$$

An interesting interpretation of relationships (2.4) is illustrated in Fig. 2.1 which represents two reliability functions such that $R_2(t) \leqslant R_1(t)$ for all t.

If the interval $(t, t + x)$ is considered, one notices that $R_2(x/t) > R_1(x/t)$ in spite of the ordering of the reliability functions. The conditional reliability during a finite time interval which does not start at $t=0$ depends on the gradient of the reliability function, not on its absolute value.

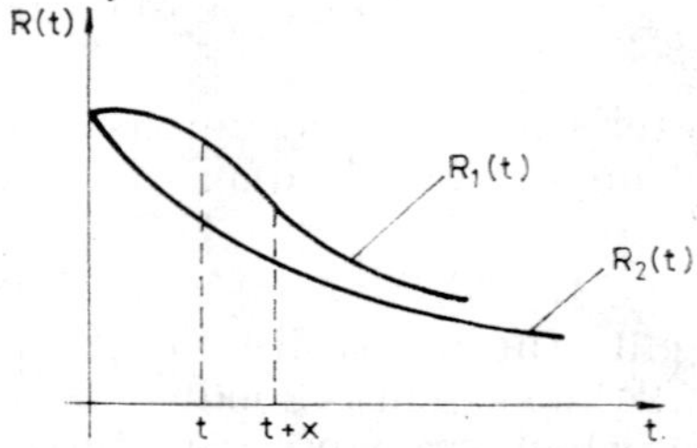

Fig. 2.1. — Interval reliability function.

The cdf and the reliability function describe the system behaviour in finite time intervals. The local behaviour of the system around a moment t is described by the probability density function (pdf) of T :

$$f(t) = \lim_{\Delta t \to 0} \frac{F(t + \Delta t) - F(t)}{\Delta t} = \frac{\mathrm{d}F(t)}{\mathrm{d}t}. \tag{2.5}$$

The pdf (2.5) has no physical interpretation but $f(t)$ multiplied by the small time interval Δt represents the total probability of failure in $(t, t+$

$+ \Delta t)$. The actual value of Δt depends on the accuracy of the model and on the measuring facilities.

In order to describe the conditional probability of failure around t, we define the failure rate $z(t)$:

$$z(t) = \lim_{\Delta t \to 0} \frac{F(t, t + \Delta t / T \geqslant t)}{\Delta t} =$$

$$= \lim_{\Delta t \to 0} \frac{F(t + \Delta t) - F(t)}{R(t) \Delta t} = \frac{f(t)}{R(t)} . \tag{2.6}$$

The failure rate $z(t)$ multiplied by the time interval Δt is the conditional probability of failure in $(t, t + \Delta t)$ given that the system has not failed before time t. From relation (2.6), it is obvious that $z(t) \geqslant f(t)$. The failure rate is analogous to the mortality rate in demographic studies and describes the instantaneous risk of system failure.

From relations (2.5) and (2.6) we get :

$$z(t) = -\frac{1}{R(t)} \frac{\mathrm{d}R(t)}{\mathrm{d}t} . \tag{2.7}$$

Integrating relation (2.7) with the initial condition $R(0) = 1$, we get :

$$R(t) = \exp \left[-\int_0^t z(u)\, \mathrm{d}u \right] \tag{2.8}$$

and, according to equation (2.4) :

$$R(x/t) = \exp \left[-\int_t^{t+x} z(u)\, \mathrm{d}u \right] . \tag{2.9}$$

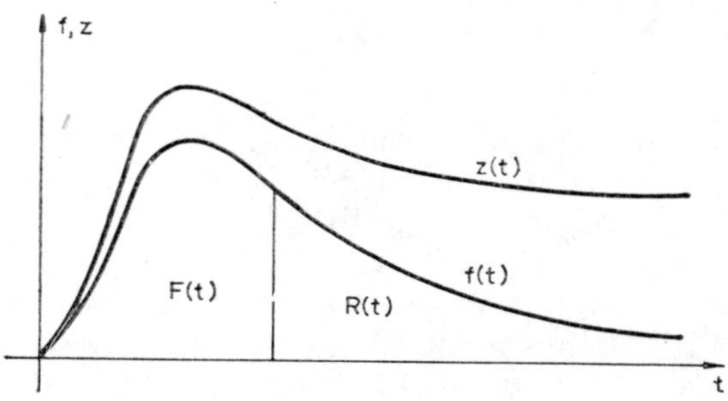

Fig. 2.2. — Reliability measures.

The relations between $R(t)$, $F(t)$, $f(t)$ and $z(t)$ are easy to establish and are shown in Table 2.1 and Fig. 2.2.

Apart from the afore-mentioned functions which describe the evolution of a system towards failure, the system reliability may be expressed

Table 2.1

The inter-relationships between the reliability measures

	$F(t)$	$R(t)$	$f(t)$	$z(t)$
$F(t)$	1	$1 - R(t)$	$\int_0^t f(u)\,du$	$1 - e^{-\int_0^t z(u)\,du}$
$R(t)$	$1 - F(t)$	1	$\int_t^\infty f(u)\,du$	$e^{-\int_0^t z(u)\,du}$
$f(t)$	$\frac{dF(t)}{dt}$	$-\frac{dR(t)}{dt}$	1	$z(t)R(t)$
$z(t)$	$\frac{1}{1-F(t)}\frac{dF(t)}{dt}$	$-\frac{1}{R(t)}\frac{dR(t)}{dt}$	$\frac{f(t)}{R(t)}$	1

by the moments of the time to failure. The moment of the first order is the expected value of the time to failure :

$$m = \int_0^\infty tf(t)\mathrm{d}t = \int_0^\infty R(t)\,\mathrm{d}t. \tag{2.10}$$

The expectation (2.10) represents the mean time-to-failure (MTTF) for non-repairable systems. In the case of repairable systems, m may be interpreted either as the mean time to the first failure (MTTFF) or as the mean time between (successive) failures (MTBF). In the latter case, the MTBF is unique only if the system recovers to the same state after each failure. Otherwise, the MTBF values must be computed using different reliability functions for different time intervals between failures. It is possible that, after each failure, the system recovers its initial state. In this case the MTTFF is equal to the MTBF.

The expected time-to-failure predicts the future failure of a system given that it is in the initial state at the prediction time. It is also useful to predict the future failure at an arbitrary moment t, given that the system is operational at t, that is the system age is t. This conditional expected value is the mean residual life (MRL):

$$m(t) = \int_0^\infty R(x/t)\,\mathrm{d}x = \frac{1}{R(t)}\int_t^\infty R(u)\,\mathrm{d}u. \tag{2.11}$$

A graphical interpretation of m and $m(t)$ is given in Fig. 2.3.

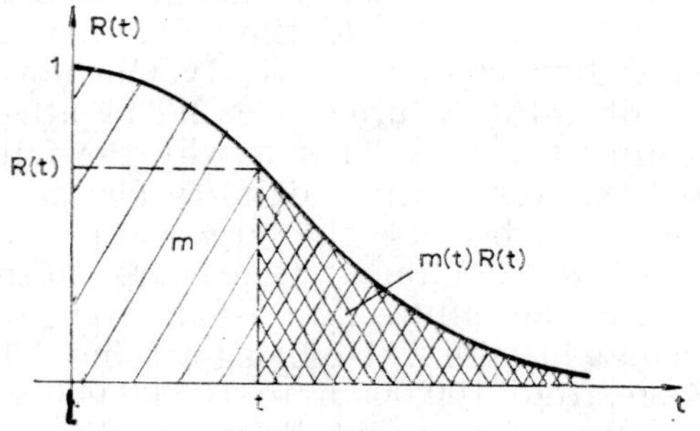

Fig. 2.3. — Mean time to failure and mean residual life.

A measure of the similarity between the members of a class of systems is provided by the central moment of the second order, that is the variance of the time-to-failure:

$$V = \int_0^\infty (t - m)^2 f(t)\,\mathrm{d}t = \int_0^\infty t^2 f(t)\,\mathrm{d}t - m^2. \tag{2.12}$$

The standard deviation $\sigma = \sqrt{V}$ is often used together with the variance (2.12). The uncertainty about the reliability behaviour of a system is reduced when the values of V and σ are small. This is a feature of the systems developed under closely controlled technological conditions.

Another component of the general reliability model is the α-percentile of the time-to-failure. This is the solution of the equation:

$$F(t_\alpha) = \alpha. \tag{2.13}$$

The α-percentile may be interpreted as a warranty time during which the fraction of failures from a population is about the specified α.

The above model, based on the mathematical characteristics of the time-to-failure, is a classical statistical approach to the reliability problem.

A selection must be made from the reliability measures derived from this model in order to describe realistically the reliability of a certain system. When the duration of operation of a system is specified, the most adequate reliability measure is the reliability function or the probability of failure during the operation period. Otherwise, one may use the reliability function for an arbitrary time interval, say 1000 hours. However, it is preferable in such cases to use the MTTF or the failure rate $z(t)$ because of their greater sensitivity to changes in the system reliability. From equations (2.8) and (2.10) it may be seen that large variations in $z(t)$ and m lead to comparatively small variations in the reliability function. For instance, let the failure rate $z(t)$ be constant and equal to $10^{-5}h^{-1}$, that is 1%/1000 h. From relations (2.8) and (2.10) we get $m = 100\,000$ h and $R(1000) = 0.99$. If the system reliability changes so that the failure rate becomes $10^{-6}h^{-1}$, that is 0.1%/1000 h, the reliability function and the MTTF will be 0.999 and 1000 000 h respectively. One notices that a change of an order of magnitude in the failure rate or MTTF affects only the third decimal of the reliability function. That is why the failure rate and the MTTF are often used to express the reliability characteristics. Of them, the failure rate is preferred because the large values of MTTF could be misleading when related to usual human time spans. In the case of electronic devices, for instance, the failure rate usually ranges between $10^{-5}h^{-1}$ and $10^{-7}h^{-1}$, levels such as $10^{-10}h^{-1}$ also being attained. The corresponding values of MTTF range from 100 000 h to 10 000 000 h and even 10^{10} h. These are so large in human terms that distinction between the reliability levels in terms of MTTF becomes difficult. The values of the failure rate are better distinguished, because they are not related to other forms of human experience.

The reliability measures derived from the general model are interdependent but this does not imply that the system reliability could be satisfactorily described on the basis of a single measure [2]. Moreover,

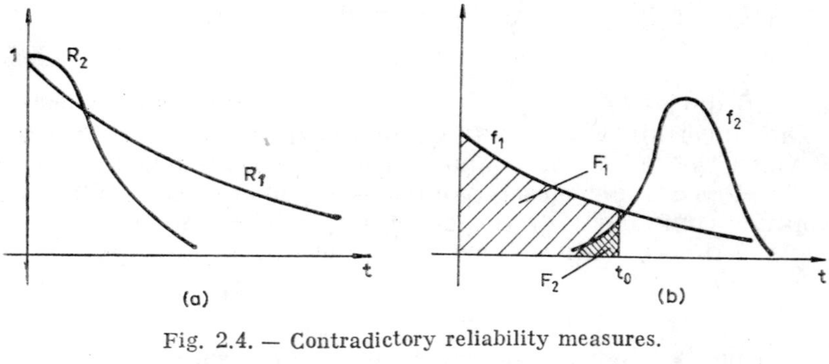

Fig. 2.4. — Contradictory reliability measures.

when comparing two systems, the reliability measures could be contradictory, so that a decision must be based on the entire set of measures. For instance, in Fig. 2.4, two reliability functions have the property :

$$\begin{aligned} &R_2(t) > R_1(t), \qquad t < t_0 ; \\ &R_2(t) \leqslant R_1(t), \qquad t \geqslant t_0. \end{aligned} \tag{2.14}$$

On the other hand,

$$m_2 = \int_0^\infty R_2(t)\,\mathrm{d}t < m_1 = \int_0^\infty R_1(t)\,\mathrm{d}t. \tag{2.15}$$

If the comparison were based solely on the MTTF values, the first system would seem superior. But, according to relation (2.14), the second system is preferable whenever the operation period is less than t_0. From the pdfs of the systems, depicted in Fig. 2.4(b), one can see that the variance of the second system is smaller, so that its behaviour is more predictable. The first system, although it is better in expectation, behaves less satisfactorily for $t < t_0$ because of its large variance of the time-to-failure. The decision on which system is better must be taken according to the values of the MTTFs, of the variances and of the reliability functions, correlated with the characteristics of the operation and the conditions of use.

The general reliability model is based on the distribution law of the time-to-failure. The distribution law must have such properties as to describe realistically the aging process of the system.

2.2 Modelling of the aging process

A system ages when its reliability characteristics change. In the life cycle of any system one may distinguish three stages which are analogous to the stages of human life. In the early youth of a system, age causes the reliability characteristics to improve. This is the period of infant mortality, when the reliability improvement is explained either by the elimination of the weak members of a collection of similar systems or by early interventions for repairing the system.

The next stage is the maturity period during which the reliability characteristics do not alter. This is the useful life of the system during which it does not age. Finally, the wearing-out period is characterized by decay of the system reliability characteristics.

The three stages differ, in their length and importance, from a system to another. A classification of systems based on their different aging processes is carried out in [1], taking into account the evolution of their reliability characteristics.

Definition 1. A system is IFR (with an increasing failure rate) if the conditional reliability function $R(x/t)$ decreases with t for any $x \geqslant 0$. For IFR systems, the reliability decreases with age. Conversely, a system is DFR (decreasing failure rate) if $R(x/t)$ increases with t for any $x \geqslant 0$. The reliability of a DFR system improves with its age. The IFR or DFR character of a system may be equivalently defined using its failure rate.

From relation (2.6), we may write

$$z(t) = \lim_{x \to 0} \frac{R(x/t)}{x} \cdot \tag{2.16}$$

The failure rate is then increasing for IFR and decreasing for DFR systems.

Conversely, the conditional reliability function may be written :

$$R(x/t) = \exp\left[-\int_t^{t+x} z(u)\,\mathrm{d}u\right] \cdot \tag{2.17}$$

It follows from relation (2.17) that, if $z(t)$ increases with t, the absolute value of the exponent increases and $R(x/t)$ decreases with t. When $z(t)$ decreases, the conditional reliability function $R(x/t)$ increases. The IFR or DFR character may be defined either on the basis of the conditional reliability function $R(x/t)$ or on the basis of the failure rate $z(t)$.

Definition 2. This describes a less strong process than IFR or DFR. A system is IFRA (increasing failure rate average) if the function $(1/t)\ln(1/R)$ is increasing. Conversely, a system is DFRA (decreasing failure rate average) if the function $(1/t)\ln(1/R)$ is decreasing. It can be proved that any IFR (DFR) system is also IFRA (DFRA). From equation (2.8), we may write :

$$\ln(1/R) = \int_0^t z(u)\,\mathrm{d}u. \tag{2.18}$$

Taking the derivative of (2.18) we get :

$$(\ln(1/R))'_t = z(t). \tag{2.19}$$

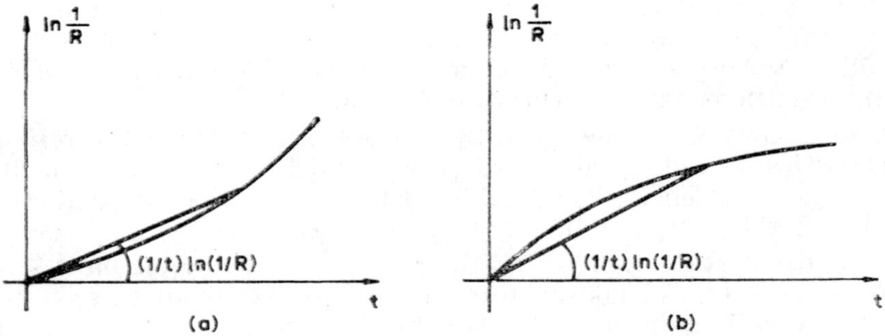

Fig. 2.5. — (a) An IFR system is IFRA ; (b) A DFR system is DFRA.

Suppose $z(t)$ is increasing. By the definition of the first derivative, the function $(1/t)\,\{\ln[1/R(t)] - \ln[1/R(0)]\} = (1/t)\ln(1/R)$ is also increasing, so that an IFR system is also IFRA. The same reasoning applies for DFR and DFRA systems.

Moreover, if $z(t)$ is differentiable, the function $\ln(1/R)$ is convex for IFR and concave for DFR systems (Fig. 2.5). If $z(t)$ is not differentiable, the function $\ln(1/R)$ is star-shaped for IFRA systems. A function $g(x)$ is star-shaped if $(1/x)g(x)$ is increasing. Any convex function which contains the origin is also star-shaped. Fig. 2.6 depicts two systems with piece-wise constant failure rates, one of which is IFR(IFRA) and the other DFR(DFRA). The function $\ln(1/R)$ in Fig. 2.6(a) is star-shaped.

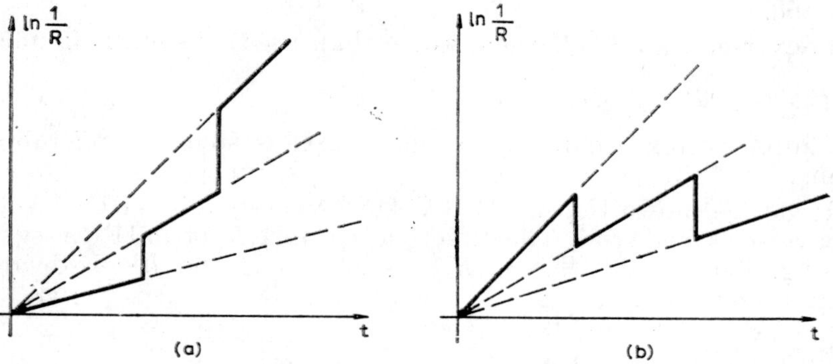

Fig. 2.6.—Piecewise constant failure rate : (a) IFR (IFRA) ; (b) DFR (DFRA).

An IFRA (DFRA) system is not necessarily IFR (DFR). It is sufficient to consider the example in Fig. 2.7. The function $\ln(1/R)$ in Fig. 2.7(a) is star-shaped, that is the system is IFRA but the failure rate $z(t)$, which is the first derivative of $\ln(1/R)$, is not an increasing function, so that the system is not IFR.

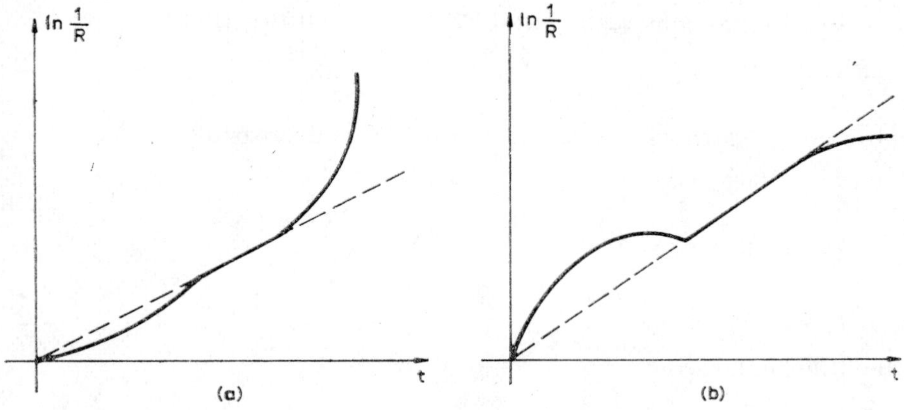

Fig. 2.7.—(a) An IFRA system which is not IFR ; (b) A DFRA system which is not DFR.

Similarly, the function $(1/t)\ln(1/R)$ in Fig. 2.7.b is decreasing, although the derivative of $\ln(1/R)$ is not, that is the DFRA system is not DFR.

Definition 3. A system is NBU (new better than used) if the conditional reliability function $R(x/t)$ is less than $R(x)$ for any age and any operating period x.

$$R(x/t) < R(x) \quad \forall x, t > 0. \tag{2.20}$$

The definition (2.20) implies that, at any age, an old system is inferior to a new one.

Conversely, an NWU (new worse than used) system is defined by :

$$R(x/t) > R(x) \quad \forall x, t > 0 \tag{2.21}$$

which implies that, whatever its age, a used system is preferable to a new one.

It can be shown that an IFRA (DFRA) system is NBU (NWU) but the converse is not true. The function $(1/t)\ \ln(1/R)$ of an IFRA system is increasing. This implies that $(1/R)^{1/t}$ is increasing and $R^{1/t}$ is decreasing. We may then write :

$$[R(t+x)]^{\frac{1}{t+x}} < [R(t)]^{\frac{1}{t}}. \tag{2.22}$$

From relations (2.22) and (2.4)

$$R(x/t) < [R(t)]^{\frac{x}{t}}. \tag{2.23}$$

Let $0 \leqslant x \leqslant t$. Then :

$$[R(t)]^{\frac{1}{t}} \leqslant [R(x)]^{\frac{1}{x}}. \tag{2.24}$$

From expressions (2.23) and (2.24), it follows that

$$R(x/t) < R(x). \tag{2.25}$$

Suppose now $x \geqslant t$. As the function $[R(t)]^{\frac{1}{t}}$ is decreasing, we have

$$[R(t+x)]^{\frac{1}{t+x}} < [R(x)]^{\frac{1}{x}} \tag{2.26}$$

which means that

$$R(t+x) < [R(x)]^{\frac{t+x}{x}}. \tag{2.27}$$

On the other hand, since $x \geqslant t$, we may write

$$[R(x)]^{\frac{1}{x}} < [R(t)]^{\frac{1}{t}} \tag{2.28}$$

which is equivalent to

$$[R(x)]^{\frac{t}{x}} < R(t). \tag{2.29}$$

From relations (2.27) and (2.29)

$$R(t + x) < R(t) \cdot R(x) \tag{2.30}$$

and consequently, $R(x/t) < R(x)$ which defines an NBU system.

In a similar way, one may prove that a DFRA system is also NWU. However, the converse is not true (the property of NBU (NWU) does not imply the IFRA (DFRA) property). An example of a system which is NBU but not IFRA is given in Fig. 2.8.

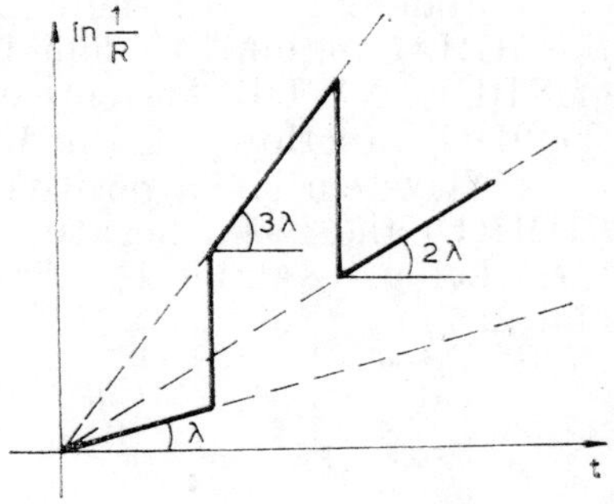

Fig. 2.8. — An NBU system which is not IFRA.

It is obvious that the system is not IFRA. In order to show the NBU property, we write the piecewise constant failure rate :

$$z(t) = \begin{cases} \lambda & t < x_0 \\ 3\lambda & x_0 \leqslant t < 2x_0 \\ 2\lambda & t \geqslant 2x_0. \end{cases} \tag{2.31}$$

Let $x_0 < x < t < 2x_0$. We have :

$$\begin{aligned} R(x) &= \mathrm{e}^{-\lambda x_0}\, \mathrm{e}^{-3\lambda(x-x_0)} = \mathrm{e}^{-\lambda(3x-2x_0)} \\ R(x/t) &= \mathrm{e}^{-3\lambda(2x_0-t)}\, \mathrm{e}^{-2\lambda(t+x-2x_0)} = \mathrm{e}^{-\lambda(2x_0-t+2x)}\ . \end{aligned} \tag{2.32}$$

As $2x_0 - t + 2x > 3x - 2x_0$, that is $t + x < 4x_0$, it follows that $R(x/t) <$ $< R(x)$ and the system is NBU.

Similar reasoning may be effected taking the values x and t in other time intervals, so the generality of the results is not affected by the particular choice of x and t.

Another definition of an aging system may be given using the mean residual life (MRL) $m(t)$.

Definition 4. A system is NBUE (new better than used in expectation) if the mean residual life at any age t is less than the MTTF, that is $m(t) < m$. Conversely, if $m(t) > m$ the system is NWUE (new worse than used in expectation). From definitions (2.10) and (2.11) of the MTTF and MRL, it follows that for an NBUE system :

$$\int_0^\infty R(x/t)\, \mathrm{d}t < \int_0^\infty R(x)\, \mathrm{d}x. \tag{2.33}$$

It is obvious from (2.33) that an NBU (NWU) system is also NBUE (NWUE). The converse is not true, for the inequality (2.33) does not imply the same inequality between the functions which are integrated. The four definitions above describe different kinds of aging processes, less and less restrictive. A summary of the implications already demonstrated is given below :

$$\text{IFR} \Rightarrow \text{IFRA} \Rightarrow \text{NBU} \Rightarrow \text{NBUE}$$

$$\text{DFR} \Rightarrow \text{DFRA} \Rightarrow \text{NWU} \Rightarrow \text{NWUE}.$$

Modelling of the aging process may also be performed in terms of the MRL function [3]. This function was already used in the definition of NBUE (NWUE) systems, which was based on the comparison between the MRL function and the MTTF.

A system has a positive memory if $m(t)$ is a decreasing function of t(DMRL) and has a negative memory if $m(t)$ is increasing with t(IMRL).

Let us examine the first derivative of $m(t)$. From equations (2.11) and (2.7)

$$m'(t) = -1 + z(t)m(t) \geqslant -1;$$

$$z(t) = \frac{1 + m'(t)}{m(t)}. \tag{2.34}$$

It is interesting to note that a limit case for a DMRL system is $m'(t) = -1$, when the system is said to have a perfect memory.

From relation (2.34) it may also be seen that the failure rate $z(t)$ is increasing when $m(t)$ is a decreasing convex function, and $z(t)$ is decreasing when $m(t)$ is an increasing concave function. A convex DMRL system is also IFR and a concave IMRL system is also DFR, but the converses are not true.

The MRL function defines a class of systems which is more restricted than the IFR (DFR) class.

A remarkable class of systems which does not fit in the above types includes those systems which do not age, that is their reliability characteristics are independent of the age of the system. Such systems have no memory because they behave at any time like new systems, regardless of how long they have been working before. The mathematical definition of a system without memory is : $R(x/t) = R(x)$, which implies that the failure rate of the system is constant. That is why a system without memory is called CFR (constant failure rate).

It may be shown [2] that there is a single distribution law of the time-to-failure which satisfies this condition, that is, the exponential distribution. Based on the exponential pdf, the components of the reliability model may be computed :

$$\begin{aligned} R(t) &= e^{-\lambda t} \\ f(t) &= \lambda e^{-\lambda t} \\ z(t) &= \lambda \\ m(t) &= m = 1/\lambda \\ V &= \sigma^2 = 1/\lambda^2 \\ t_\alpha &= (1/\lambda)\ln[1/(1-\alpha)]. \end{aligned} \tag{2.35}$$

A graphical display of the reliability measures of the exponential model is given in Fig. 2.9. Some remarkable properties of the exponential model should be pointed out. The failure rate is constant and equal to the inverse of MTTF. The MRL is constant and equal to the MTTF. The standard deviation is also equal to the MTTF.

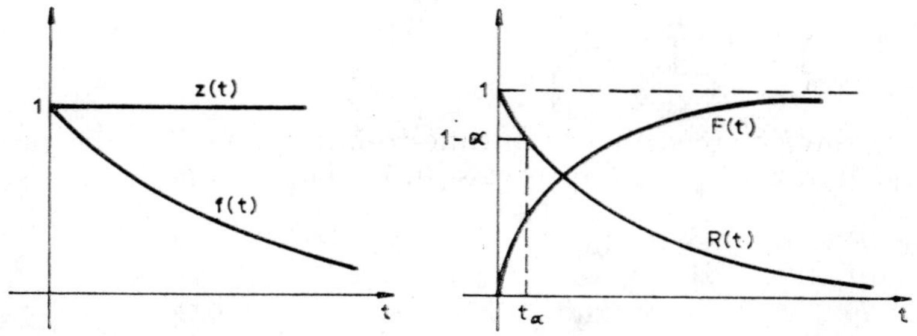

Fig. 2.9. — Reliability measures of a CFR system.

The exponential model is a limiting case because no real system is completely without memory. However, it is adequate for the maturity period of any system (when the failure rate is approximately constant). This period is more extended in the case of systems which do not use mechanical transmission such as electronic systems. The lack of memory is not always favourable from the user's point of view. The variance of the time-to-failure is large, which means a great dissimilarity among the systems of a class. This is the case of complex electronic systems which are less predictable than those based on traditional materials and simpler technologies. The systems subject to wearing-out may be initially superior with respect to the reliability function but their reliability is rapidly degraded with age. The more rapid the degradation, the smaller the uncertainty associated with the reliability prediction.

The exponential model is also theoretically important because it represents a reference case for all existing systems. Moreover, any time-to-failure distribution may be approximated by an assembly of exponential models.

It is empirically established that most of the systems are initially DFR, then CFR and finally IFR. This conclusion is depicted in the classical bath-tub curve of the failure rate. Sometimes the dependence of the reliability characteristics on age is not so strong, so it is necessary to consider models other than DFR or IFR (DFRA, IFRA and so on). It is, however, essential to establish on

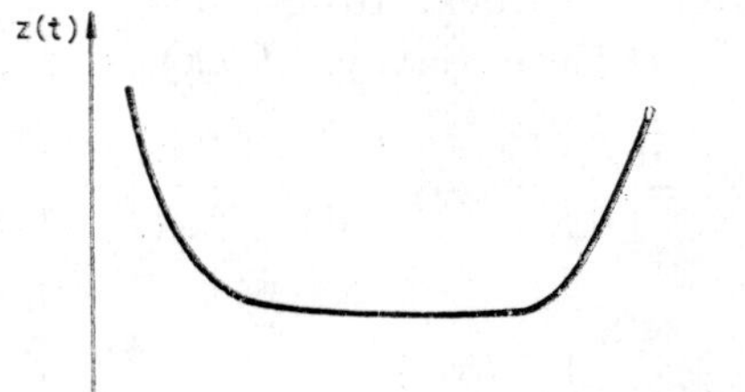

Fig. 2.10. — The bath tub curve.

an experimental basis if the failure rate of a system is constant, increasing or decreasing.

A statistical test to establish the CFR, IFR or DFR character of a system was developed in [4]. It uses the function :

$$t_\Sigma(F) = \int_0^{t_F} R(t)\,\mathrm{d}T \tag{2.36}$$

where t_F is the F-percentile of the time-to-failure. As F is the probability of failure, it ranges in the interval $[0, 1]$. The end points of $t_\Sigma(F)$ are :

$$\begin{aligned} &\text{For } F = 0,\ t_F = 0 \text{ and } t_\Sigma(F) = 0\,; \\ &\text{For } F = 1,\ t_F \to \infty \text{ and } t_\Sigma(F) = m. \end{aligned} \tag{2.37}$$

According to (2.37), the function $t_\Sigma(F)$ may be normalized, dividing it to the MTTF. We then get the function :

$$T_S(F) = \frac{t_\Sigma(F)}{t_\Sigma(1)} = \frac{1}{m}\int_0^{t_F} R(t)\,\mathrm{d}t \tag{2.38}$$

which is restricted to a unit square in the first quadrant.

We shall show that the function $T_S(F)$ is linear in the case of CFR systems, convex for DFR and concave for IFR systems. Let us consider first a CFR system. According to relation (2.35), the function $T_S(F)$ becomes :

$$T_S(F) = \frac{1}{1/\lambda}\int_0^{(1/\lambda)\ln 1/(1-F)} \mathrm{e}^{-\lambda t}\,\mathrm{d}t =$$

$$= 1 - \mathrm{e}^{-\ln 1/(1-F)} = 1 - (1 - F) = F \tag{2.39}$$

which is a linear function.

Differentiating $T_S(F)$ with respect to F, we get :

$$\frac{\mathrm{d}T_S(F)}{\mathrm{d}F} = \frac{\mathrm{d}T_S(F)}{\mathrm{d}t_F}\cdot\frac{\mathrm{d}t_F}{\mathrm{d}F} = \frac{1}{m}R(t_F)\cdot\frac{1}{\dfrac{\mathrm{d}F}{\mathrm{d}t_F}} =$$

$$= \frac{1}{m}\,\frac{R(t_F)}{f(t_F)} = \frac{1}{mz(t_F)}\,. \tag{2.40}$$

In the case of IFR systems, $z(t)$ is increasing with t. When F increases, t_F increases, so $z(t)$ increases and the first derivative $T_S(F)$ decreases, which means that $T_S(F)$ is a concave function. Conversely, in the case of DFR systems, $z(t)$ is decreasing, and the first derivative of $T_S(F)$ is increasing, so $T_S(F)$ is a convex function. These conclusions are depicted in Fig. 2.11.

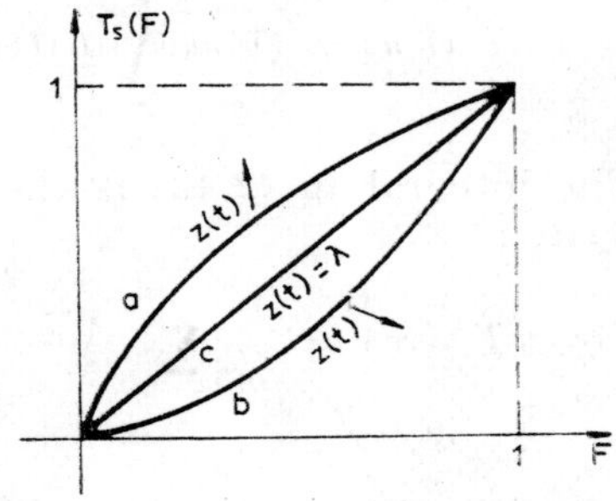

Fig. 2.11.—The Barlow-Campo test.

The test consists in estimating $T_S(F)$ from experimental data and taking a decision based on its graphical aspect. Let the data consist of an ordered sample of n times-to-failures $0 < t_1 < t_2 < \ldots < t_n$. The probability of failure $F(t)$ is estimated as the quotient of the number of failures which occurred until t by the sample number :

$$\hat{F}(t) = \begin{cases} 0 & t < t_1 \\ 1/n & t_1 \leqslant t < t_2 \\ 2/n & t_2 \leqslant t < t_3 \\ \cdots\cdots\cdots & \cdots\cdots \\ i/n & t_i \leqslant t < t_{i+1} \\ \cdots\cdots\cdots & \cdots\cdots \\ (n-1)/n & t_{n-1} \leqslant t < t_n \\ 1 & t \geqslant t_n. \end{cases} \tag{2.41}$$

The estimated probability of failure is sometimes called the empirical cdf of the time-to-failure.

The estimated reliability function is obtained by complementing (2.41), and is represented in Fig. 2.12. The function $T_S(F)$ will be estimated at the failure times t_i, where $\hat{F} = i/n$. Thus, the t_i values are the

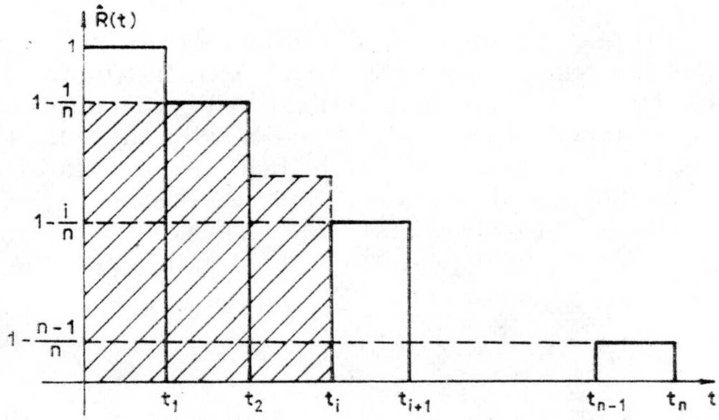

Fig. 2.12 — Total time on test estimation.

estimated i/n-percentiles of the time-to-failure. From relation (2.38), we get:

$$\hat{T}_S(i/n) = (1/\hat{m}) \int_0^{t_i} \hat{R}(t)\,\mathrm{d}t. \tag{2.42}$$

The integral in (2.42) is the shaded area in Fig. 2.11, so that we may write:

$$\hat{T}_S(i/n) = \frac{1}{\hat{m}} \sum_{k=1}^{i} \frac{n-k+1}{n} (t_k - t_{k-1})\,; \quad i = 1, 2, \ldots, n\,; \quad t_0 = 0. \tag{2.43}$$

After some reductions, (2.43) becomes:

$$\hat{T}_S(i/n) = \frac{\sum_{k=1}^{i} t_k + (n-i)\, t_i}{n\hat{m}}\,; \quad i = 1, 2, \ldots, n. \tag{2.44}$$

The numerator in (2.44) signifies the accumulated test time of the sample up to failure i, T_{Σ_i}. As the MTTF is naturally estimated by the mean of the observed failure times, the denominator of (2.44) is the sum of all the observed times-to-failure, that is the total test time of the sample, T_Σ. The estimation of T_S then becomes:

$$\hat{T}_S(i/n) = \frac{T_{\Sigma_i}}{T_\Sigma}\,; \quad i = 1, 2, \ldots, n. \tag{2.45}$$

In the case of a censored sample, when only $r \leqslant n$ times-to-failure are observed, the MTTF is estimated according to:

$$\hat{m} = \frac{T_{\Sigma_r}}{r}. \tag{2.46}$$

Relationship (2.46) may be justified as follows. When only r failures are observed, a natural estimate of m is the arithmetic mean of the r observed times-to-failure. However, this implies neglecting the information given by the items which survived the test. If we consider the total time on test and divide it by the sample n, this would lead to a pessimistic estimate, as if all the items had failed in the test. An extension of the arithmetic mean is then the quotient of the total time on test by the number of failures. The estimate of T_S, using relations (2.44) and (2.46), is then

$$\hat{T}_S(i/n) = \frac{r}{n} \frac{T_{\Sigma_i}}{T_{\Sigma_r}}\,; \quad i = 1, 2 \ldots, r. \tag{2.47}$$

The graphical representation of $\hat{T}_S$ is contained in an $(r/n \times r/n)$ square.

The decision on the IFR, DFR or CFR character of the system is drawn from the graphical representation of the estimate of $T_S(F)$. The decision may be affected by the errors which accompany any statistical test. Let us consider, for instance, a CFR system, which implies that the true $T_S(F)$ is linear. The estimate of $T_S(F)$ is normally not linear, but oscillates around the line $T_S(F) = F$, as in Fig. 2.13. The correct decision that the system is CFR may be reached in this case by a simple inspection of the graph, but a quantitative decision rule may also be employed [4].

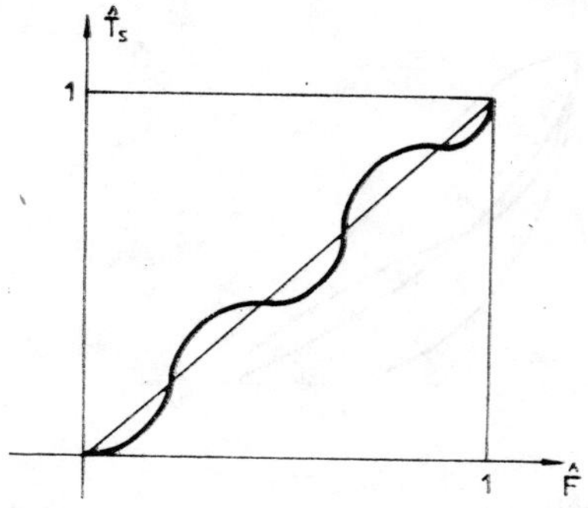

Fig. 2.13. — Barlow-Campo test for a CFR system.

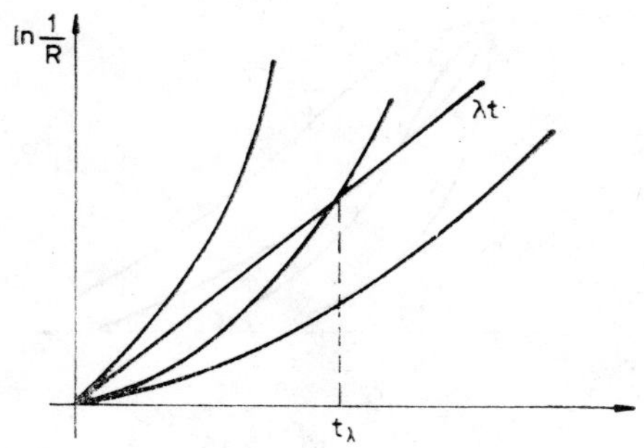

Fig. 2.14. — Properties of the reliability function.

According to this rule, the system is considered CFR if the number of intersections of $\hat{T}_S(F)$ and the first bisector exceeds a certain limit, which depends on the sample number and on the risk of the first order. However, evaluation of the power of the test is difficult, and must rely mainly on simulation methods.

The above considerations on the characteristics of the aging process help to establish the general form of the reliability function for different classes of systems. It was shown that $\ln(1/R)$ is a convex (or star-shaped) function in the case of IFRA systems and is linear in the case of CFR systems. This implies that $\ln(1/R)$ and λt have at most one intersecting point besides the origin. If $\ln(1/R)$ is asymptotically parallel to λt, the intersecting point is shifted to infinity (see Fig. 2.14), while if $z(0) > \lambda$ the only intersection is at $t = 0$.

Thus, the reliability function of an IFRA system intersects the exponential reliability function of a CFR system at most once, so that it is above the exponential before the intersection and then it goes below it. If there is no intersection, the reliability function of the IFRA system may be either above or below the exponential. A similar reasoning shows that the reliability function of a DFRA system intersects the exponential reliability function of a CFR system at most once, so that it is below the exponential before the intersection and then it goes above. These considerations are depicted in Fig. 2.15 where the three possible shapes of the reliability function with respect to the exponential are identified by I, II and III.

Considering the variable $x = e^{-\lambda t}$ instead of t, the shapes of the reliability functions are more clearly outlined (Fig. 2.16). Both the IFRA and the DFRA systems may have convex (I) or concave (III) reliability functions with respect to the exponential. The S-shaped reliability function II shows that an IFRA system may be superior to a CFR system for short operating periods and inferior for longer ones.

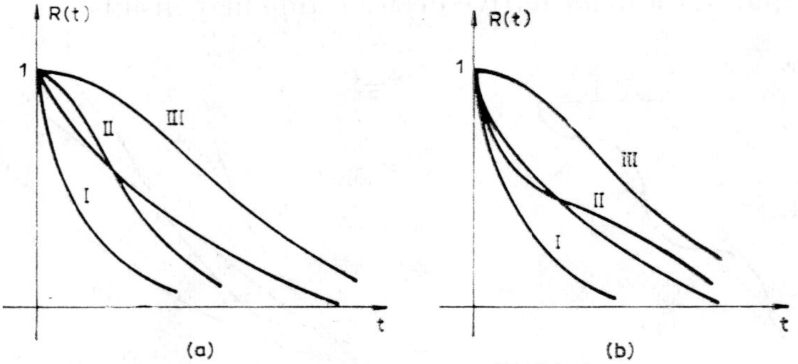

Fig. 2.15. — Reliability function vs time : (a) IFRA system ; (b) DFRA system.

The reverse is true in the case of the inverted S-shaped reliability function of the DFRA systems.

The general form of the reliability function is useful in establishing margins for the system reliability.

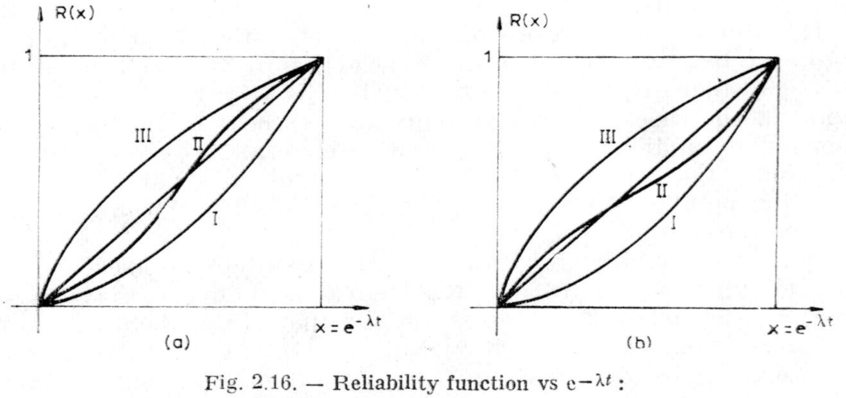

Fig. 2.16. — Reliability function vs $e^{-\lambda t}$: (a) IFRA system ; (b) DFRA system.

Suppose that the value of the reliability of an IFRA system for a specified operating time t_0 is $R_0 = R(t_0)$. A lower limit of the reliability function for $t < t_0$ may be established as follows. A CFR system is determined so that it intersects $R(t)$ at (t_0, R_0). Its failure rate satisfies the relationship :

$$R_0 = e^{-\lambda t_0};$$
$$\lambda = \frac{1}{t_0} \ln \frac{1}{R_0}. \quad (2.48)$$

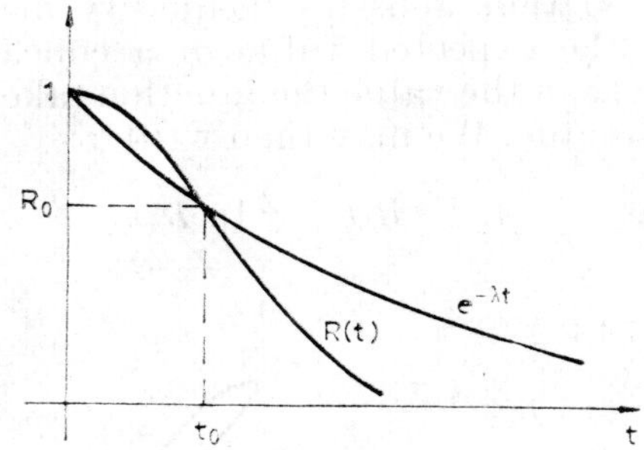

Fig. 2.17. — Limits for the reliability function when $R_0 = R(t_0)$ is known.

The exponential $e^{-\lambda t}$ is a lower limit for $R(t)$ $(t < t_0)$ because of the IFRA property of the system (see Fig. 2.17). We may then write

$$R(t) > e^{-\lambda t} = e^{-(t/t_0)\ln(1/R_0)}. \quad (2.49)$$

For periods longer than t_0, $e^{-\lambda t}$ is an upper limit of $R(t)$.

In the case of a DFRA system with a known particular value $R_0 = R(t_0)$ of the reliability function, the exponential function which intersects $R(t)$ at (t_0, R_0) is an upper limit of $R(t)$ for $t < t_0$ and a lower limit for $t > t_0$.

Suppose now that, instead of a particular value of the reliability function, the MTTF of an IFRA system is known. We shall derive a limit of the reliability function $R(t)$ for $t > m$. Consider a CFR system which is discarded at time t irrespective whether it has failed or not, and assume that its MTTF is equal to the MTTF of the IFRA system. We may then write :

$$\int_0^t e^{-\lambda t}\, dt = \int_0^\infty R(t)\, dt = m. \quad (2.50)$$

The solution of (2.50), which exists only for $t > m$, may be derived from the equation :

$$\frac{1}{\lambda}(1 - e^{-\lambda t}) = m. \quad (2.51)$$

Equality (2.50) implies that the reliability function $R(t)$ and $e^{-\lambda t}$ have an intersection point in time interval $(0, t)$, according to Fig. 2.18. Consequently, an upper limit of $R(t)$ is $e^{-\lambda t}$, where λ is the solution of (2.51).

In the case of a DFRA system the same reasoning holds except that $e^{-\lambda t}$ is a lower limit of $R(t)$.

A limit of the reliability function for $t < m$ may be derived only for an IFR (DFR) system with known MTTF. Assume that the failure rate of the IFR system is differentiable. Then, $\ln R$ is a concave function, so that Jensen's inequality may be applied. This inequality states that the expected value of a concave function of a random variable is less than the value the function takes when the variable is equal to its expected value. We may then write :

$$E[\ln R(t)] \leqslant \ln R(m). \tag{2.52}$$

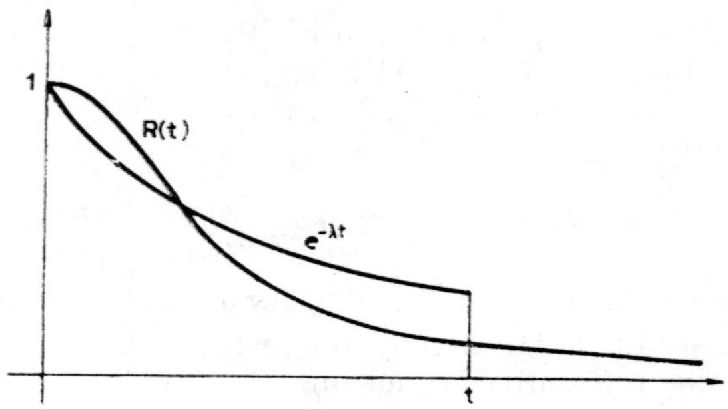

Fig. 2.18. — Limits for the reliability function when the MTTF is known.

The expected value of $\ln[R(t)]$ may be computed directly :

$$E[\ln R(t)] = \int_0^\infty [\ln R(t)] f(t)\,\mathrm{d}t = \int_0^1 \ln R\,\mathrm{d}R = -1. \tag{2.53}$$

From (2.52) and (2.53) it follows that :

$$\ln R(m) \geqslant -1 ; \tag{2.54}$$

$$R(m) \geqslant \mathrm{e}^{-1}.$$

An IFR system is also IFR A, so that $(1/t)\ln(1/R)$ is increasing and $[R(t)]^{1/t}$ is a decreasing function. Thus, for $t < m$ we may write

$$[R(t)]^{1/t} \geqslant [R(m)]^{1/m} \geqslant \mathrm{e}^{-1/m} ; \tag{2.55}$$

$$R(t) \geqslant \mathrm{e}^{-t/m}.$$

According to (2.55), $\mathrm{e}^{-t/m}$ is a lower limit of $R(t)$ for $t < m$. This limit may not be used for an IFRA system which is not IFR, because for such a system the function $\ln R(t)$ is not necessarily concave.

In the case of DFR systems, similar reasoning shows that $\mathrm{e}^{-t/m}$ is an upper limit of $R(t)$ for $t < m$.

The limits of the reliability function emphasize the importance of the exponential distribution in the modelling of failure mechanisms. However, the exponential law alone cannot adequately describe the evolution of systems subject to an aging process.

2.3 Distribution laws associated with failure mechanisms

A central problem of reliability theory is the derivation of an analytical expression for the reliability function. The distribution laws developed from the classical statistical approach are adopted in the reliability field to the extent that they provide a suitable form for the reliability function. The adoption of distribution laws from other fields, such as economics or demography in order to describe the failure mechanisms, seems an artificial approach which lacks insight into the physical phenomena.

That is why much research has been carried out in order to derive, on the basis of the physics of failure, a distribution law strictly appropriate to a certain failure mechanism. A slightly less ambitious goal is the association of physical interpretations with the theoretical distributions available, so that the application area of a distribution law should be better defined. Nevertheless, the purely physical approach has led to rather modest achievements. The problem of deriving a distribution law starting from physical considerations was stated in [9]. The distribution law which governs the degradation process of a pn junction was derived in [7] but the result is difficult to use. Research in the field of integrated circuits showed that physical reasoning may help to reduce the effort of life testing, but cannot by itself lead to a complete description of the failure process [8].

Combining physical analysis with experimental data, some distribution laws specific to the reliability field have been derived, for example those of Birnbaum and Saunders, Gumbel, Weibull and the alpha distribution. On the other hand intuitive physical interpretations, associated with the classical distribution laws of Gauss, Rayleigh or the so-called lognormal, suggest possible applications in reliability [6].

The classical reliability textbooks [5, 10] present in detail the properties of the distribution laws currently employed in reliability. Other distributions, which have recently been adopted, are discussed in [11]. We shall restrict ourselves to a presentation of the more widely used distribution laws in Table 2.2, which contains the pdf, expectation and variance of each distribution.

The association of a distribution law with a specific failure mechanism must be sustained by the physical interpretation and the experimental data (which have the last word). As the derivation of a distribution law specific to a certain mechanism is almost impossible, the usual course is to select one among the available distribution laws and test it against the experimental data.

The following example [5, 6] shows how a physical interpretation may help to select the right distribution law.

Consider a wafer which has a surface defect, of an initial random depth X_0. Let X_i be the depth of the defect at time t_i and assume that failure occurs when the critical value X_n is reached. Assume

Table 2.2
Time-to-failure distributions

$f(t)$	Name	m	V
$\dfrac{\beta(\beta t)^{\alpha-1}\,e^{-\beta t}}{\Gamma(\alpha)}$	gamma	$\dfrac{\alpha}{\beta}$	$\dfrac{\alpha}{\beta^2}$
$\dfrac{\beta t^{\beta-1}}{\alpha}\,e^{-\frac{t^\beta}{\alpha}}$	Weibull	$\sqrt[\beta]{\alpha}\,\Gamma\left(1+\dfrac{1}{\beta}\right)$	$\alpha^{\frac{2}{\beta}}\left[\Gamma\left(1+\dfrac{2}{\beta}\right)-\Gamma^2\left(1+\dfrac{1}{\beta}\right)\right]$
$\dfrac{1}{\sqrt{2\pi}\sigma}\,e^{-\frac{1}{2}\left(\frac{t-\mu}{\sigma}\right)^2}$	Gauss	μ	σ^e
$\dfrac{1}{\sqrt{2\pi}\sigma t}\,e^{-\frac{1}{2}\left(\frac{\ln t-\mu}{\sigma}\right)^2}$	lognormal	$e^{\mu+\frac{\sigma^2}{2}}$	$e^{2\mu+\sigma^2}(e^{\sigma^2}-1)$
$\dfrac{2}{\omega^2}\,t\,e^{-t^2/\omega^2}$	Rayleigh	$\omega\Gamma\left(\dfrac{3}{2}\right)$	$\left[1-\Gamma^2\left(\dfrac{3}{2}\right)\right]\omega^2$
$\dfrac{2\theta^{k+1}}{\Gamma(k+1)}\,t^{2k+1}\,e^{-\theta t^2}$	Generalized Rayleigh (GRV)	$\dfrac{\Gamma\left(k+\frac{3}{2}\right)}{\Gamma(k+1)}\,\dfrac{1}{\sqrt{\theta}}$	$\left[k+1-\dfrac{\Gamma^2\left(k+\frac{3}{2}\right)}{\Gamma^2(k+1)}\right]\dfrac{1}{\theta}$
$\dfrac{\beta}{t^2\sqrt{2\pi}}\quad{}^{-\frac{1}{2}\left(\frac{\beta}{t}-\alpha\right)^2}$	alpha	$\dfrac{\beta}{\alpha}\left(1+\dfrac{1}{\alpha^2}\right)$	$\dfrac{\beta^2}{\alpha^4}\left(1+\dfrac{8}{\alpha^2}\right)$
$\delta b^{-\varepsilon}t^{\delta-1}$ $0 < t < b$	power	$\dfrac{\delta b}{\delta+1}$	$\dfrac{\delta b^2}{(\delta+2)(\delta+1)^2}$
$\dfrac{1}{2\sqrt{2\pi\alpha^2\beta t^2}}\,\dfrac{t^2-\beta^2}{(t/\beta)^{1/2}-(\beta/t)^{1/2}}\times$ $\times\,e^{-\frac{1}{2\alpha^2}\left(\frac{t}{\beta}+\frac{\beta}{t}-2\right)}$	Birnbaum Saunders	$\beta\left(1+\dfrac{\alpha^2}{2}\right)$	$(\alpha\beta)^2\left(1+\dfrac{5\alpha^2}{4}\right)$

also that the increment $X_i - X_{i-1}$ of the defect is proportional to X_{i-1}:

$$X_i - X_{i-1} = \pi_i X_{i-1}, \tag{2.56}$$

where π_i are also random values.

From (2.56),

$$\sum_{i=1}^{n} \pi_i = \sum_{i=1}^{n} \frac{X_i - X_{i-1}}{X_{i-1}}. \tag{2.57}$$

If the increment $\Delta X_{i-1} = X_i - X_{i-1}$ is small and the number of stages n is great, (2.57) becomes:

$$\sum_{i=1}^{n} \pi_i = \int_{X_0}^{X_n} \frac{\mathrm{d}x}{x} = \ln X_n - \ln X_0. \tag{2.58}$$

Relation (2.58) may be written, equivalently:

$$\ln X_n = \ln X_0 + \sum_{i=1}^{n} \pi_i. \tag{2.59}$$

The logarithm of the critical defect is equal to the sum of a great number of random variables. According to the central limit theorem, $\ln X_n$ is asymptotically normally distributed, that is X_n has a lognormal distribution. As the time to failure is proportional to X_n, we may adopt a lognormal distribution as the reliability model. This model should then be tested against experimental data in order to validate it and to specify its parameters.

The essential step in adopting a distribution law is a goodness-of-fit test based on the theory of hypothesis testing. According to this theory, a null hypothesis H_0 is formulated by selecting a possible candidate for the life distribution among the available distribution laws, by physical reasoning. The alternate hypothesis H_1 includes all distributions except the one selected by the null hypothesis. A decision is taken according to a goodness-of-fit test and is prone to two types of errors described by the probability α of rejecting the null hypothesis when it is true and by the probability β of accepting the null hypothesis when it is false. The risks α and β must not exceed some pre-determined values upon which the decision rule is based.

The best known goodness-of-fit test is the Kolmogorov-Smirnov test which uses the times-to-failure of all the units of a sample [12]. Let $\{t_i;\ i = 1, 2, \ldots, n\}$ be the ordered times-to-failure. The cdf $F(t)$ may be estimated from the t_i values, using relation (2.41). The Kolmogorov-Smirnov theorem states that the maximum deviation of the estimated cdf from the true cdf, that is $\sup_t |\hat{F}(t) - F(t)|$ is a random variable governed by a distribution law which is independent of $F(t)$. This distribution law, called the Kolmogorov-Smirnov law, depends only on the sample n and is extensively tabulated. The decision rule is:

Accept if and only if (iff) :

$$\sup_t |F(t) - \hat{F}(t)| < e_{1-\alpha}(n) \tag{2.60}$$

where $e_{1-\alpha}(n)$ is the $1-\alpha$ percentile of the Kolmogorov-Smirnov distribution. It is clear that, if the null hypothesis is correct, the probability of rejecting it is α. Thus, the decision rule is determined by the risk of the first order. It is difficult to derive the risk of the second order β because of the very way the hypotheses are formulated. For a precise alternate distribution, the risk β may be estimated by simulation.

In practical applications, the parameters of the hypothesized distribution are unknown and, therefore, must be estimated from the data. The most common procedure is the least squares estimation. This procedure uses probability charts, with a particular choice of the quantities on the coordinate axes such that the theoretical cdf should be represented as a straight line. The estimated cdf $\hat{F}(t)$ is depicted in the reliability chart together with the line which best approximates the points $[t_i, \hat{F}(t_i)]$. This line represents the theoretical cdf and its parameters may readily be estimated graphically. The maximum deviation between the estimated cdf $\hat{F}(t)$ and the true cdf $F(t)$ is compared to the $1 - \alpha$ percentile of the Kolmogorov-Smirnov distribution and the decision rule (2.60) is applied. Let us consider as an example the Weibull distribution :

$$F(t) = 1 - e^{-t^{\beta}/\alpha}. \tag{2.61}$$

A logarithmic transformation of (2.61) leads to :

$$\ln\ln \frac{1}{1 - F} = \beta \ln t - \ln \alpha. \tag{2.62}$$

The probability chart for the Weibull distribution is obtained taking the unit quantity on the $F(t)$ axis proportional to $\ln\ln 1/(1 - F)$ and that on the t axis proportional to $\ln t$ [13, 14]. The data are plotted on the probability chart and the line which best approximates the data is traced, using the method of least squares. This line is the theoretical cdf, selected from the family distribution as the one which is best fitted to the observed data.

The procedure described above is not entirely correct because the classical Kolmogorov-Smirnov test and the related tables of percentiles are not designed for a case when the parameters of the hypothesized distribution are derived from the sample data. The general method still holds in this case, but the decision rule must be modified as follows. Once the parameters are estimated, a Monte Carlo simulation with the same sample size as the data should be performed in order to estimate the cdf of the maximum deviation. The limit of (2.60) is then determined as a percentile from this empirical cdf, not from the classical Kolmogorov-Smirnov distribution [15].

A disadvantage of the classical Kolmogorov-Smirnov test is that it uses a complete sample, that is all the times-to-failure have to be recorded. The test was adapted for censored samples, and adequate tables for the decision rule have been provided [16]. Apart from the Kolmogorov-Smirnov test, analytical and graphical tests like the χ^2 test, the binomial test and the total time test are also used [17]. A general drawback of all these tests is that their selective power is small for the usual sample sizes. Almost any hypothesis is then difficult to reject. This fact is illustrated in [17] where three hypotheses on the distribution law (exponential, Weibull and normal) are tested for the same sample, of size 10, and all of them are accepted, the risk α of rejecting a true hypothesis being set at 10%.

As the risk β of accepting a false hypothesis is not specified, the decision to reject a hypothesis is more significant than the decision to accept it. The significance level is equal to $1 - \alpha$, that is the probability of accepting a true hypothesis, which is initially specified. That is why it is necessary to test various hypotheses against the same sample data, adopting finally the one that has not been rejected by the test. If several hypotheses have passed the test, the selection must be based on other than statistical considerations.

Poor discrimination is not the only disadvantage of the goodness-of-fit tests. Once a hypothesis is accepted, it is tacitly assumed that the distribution law is also valid outside the time interval in which the test was performed. It is quite possible, however, that the distribution law should change, in a way the goodness-of-fit test could not predict.

A new approach to the selection of a theoretical reliability model is offered by the concept of informational energy [18, 19].

Let S and T be two discrete distribution laws which associate with the same random variable the probabilities $(p_1, p_2, \ldots, p_n)$ and $(q_1, q_2, \ldots, q_n)$ respectively, such that $\sum_{i=1}^{n} p_i = \sum_{i=1}^{n} q_i = 1$. The informational correlation between the two probability laws is:

$$C_{S,T} = \sum_{i=1}^{n} p_i q_i. \tag{2.63}$$

The correlation between two identical probability laws is called the informational energy

$$\begin{aligned} C_{S,S} &= E_S = \sum_{i=1}^{n} p_i^2 \\ C_{T,T} &= E_T = \sum_{i=1}^{n} q_i^2. \end{aligned} \tag{2.64}$$

The coefficient of informational correlation is defined by:

$$R_{S,T} = \frac{C_{S,T}}{\sqrt{E_S E_T}}. \tag{2.65}$$

It is obvious, from (2.65), that the coefficient $R_{S,T}$ is at most equal to unity in the case of two identical probability laws.

Now, let S be an empirical cdf, estimated from sample data, and T be the theoretical cdf. If the theoretical model is right, the coefficient $R_{S,T}$ tends to unity as the sample size increases. For a given sample size, $R_{S,T}$ is a measure of adequacy of the theoretical model for the data. A goodness-of-fit test, based on the value of this coefficient may be designed.

Another information theory approach to reliability modelling is the maximum entropy principle. The probabilistic law $f(x)$ associated with a random phenomenon must be strictly compatible with the information available on the phenomenon, and it must not in any way underrate the uncertainty about the possible outcomes. A measure of this uncertainty is the entropy, defined as:

$$H = -\int_0^\infty f(x) \log f(x)\, \mathrm{d}x. \tag{2.66}$$

The principle of maximum entropy states that the distribution law $f(x)$ associated with a physical random variable should be selected so that the entropy (2.66) is maximized, subject to the constraints imposed by the experimental data. When no data are available, the maximization of (2.66) leads to a uniform distribution, which implies that nothing is known about the phenomenon. Suppose that some information is available, and that it consists of the expectations C_i of k functions $g_i(x)$ of the random variable to be modelled:

$$C_i = \int_0^\infty g_i(x)\, f(x)\, \mathrm{d}x\,; \quad i = 1, 2, \ldots, k. \tag{2.67}$$

Apart from the constraints (2.67), the pdf $f(x)$ must satisfy the conditions:

$$f(x) \geqslant 0\,;$$

$$\int_0^\infty f(x)\, \mathrm{d}x = 1. \tag{2.68}$$

We shall apply the method of Lagrange multipliers in order to derive $f(x)$. For simplicity, we assume $k = 1$. A linear combination of (2.66)–(2.68) is

$$H - \alpha - \beta C = \int_0^\infty f(x) \log \frac{\mathrm{e}^{-[\alpha + \beta g(x)]}}{f(x)}\, \mathrm{d}x. \tag{2.69}$$

According to the inequality $\log x \leqslant x - 1$ for any $x \geqslant 1$, we may write

$$H - \alpha - \beta C \leqslant \int_0^\infty f(x) \left[\frac{e^{-[\alpha + \beta g(x)]}}{f(x)} - 1 \right] dx =$$

$$= e^{-\alpha} \int_0^\infty e^{-\beta g(x)} \, dx - 1. \tag{2.70}$$

The right hand side of (2.70) is equal to zero for :

$$\alpha = \ln \int_0^\infty e^{-\beta g(x)} \, dx. \tag{2.71}$$

According to equation (2.71), (2.70) becomes :

$$H \leqslant \ln \int_0^\infty e^{-\beta g(x)} \, dx + \beta C. \tag{2.72}$$

The entropy is maximum for :

$$f(x) = e^{-[\alpha + \beta g(x)]} = \frac{e^{-\beta g(x)}}{\int_0^\infty e^{-\beta g(x)}} . \tag{2.73}$$

The constant β is determined from (2.67) :

$$C = \int_0^\infty g(x) f(x) \, dx = \frac{\int_0^\infty g(x) e^{-\beta g(x)} \, dx}{\int_0^\infty e^{-\beta g(x)} \, dx}. \tag{2.74}$$

Relation (2.74) is equivalent to :

$$\frac{\partial [\ln Z(\beta)]}{\partial \beta} = - C \tag{2.75}$$

where :

$$Z(\beta) = \int_0^\infty e^{-\beta g(x)} \, dx.$$

These results may readily be generalized for the case when the expectations of k functions $g_1(x), g_2(x), \ldots, g_k(x)$ are known. The pdf $f(x)$ which maximizes the entropy is

$$f(x) = \frac{e^{-\sum_{i=1}^{k} \beta_i g_i(x)}}{\int_0^\infty e^{-\sum_{i=1}^{k} \beta_i g_i(x)} dx}. \tag{2.76}$$

The constants $\beta_1, \beta_2, \ldots, \beta_k$ may be determined from the equations :

$$\frac{\partial}{\partial \beta_i} [\ln Z(\beta_1, \beta_2, \ldots, \beta_k)] = - C_i; \quad i = 1, 2, \ldots, k$$

where

$$Z(\beta_1, \beta_2, \ldots, \beta_k) = \int_0^\infty e^{-\sum_{i=1}^{k} \beta_i g_i(x)} dx. \tag{2.77}$$

In practical situations the expectations C_i are not known, but may be estimated as the sample averages :

$$C_i = \frac{1}{n} \sum_{i=1}^{n} g_i(x_i). \tag{2.78}$$

The functions $g_i(x)$, which are essential in building up the distribution law, are unknown, but can be selected from a larger set of potential information functions. The selection is based upon the random sample $(x_1, x_2, \ldots, x_n)$. It is sufficient to consider the problem as a pdf approximation on a bounded interval $[a, b]$, since one seeks to fit a pdf to a random sample which is bounded by the smallest and largest order statistics. If the data come from an unbounded random variable having finite variance and the sample size is moderate to large ($n \geqslant 50$), then the approximation will still be good [21]. A good starting potential set of information functions is given in Table 2.3. The selection of an active set among these functions is necessary, in order to use as few information functions in the entropy approximation as possible. The active set is selected by minimizing the information exchange between the pdf of maximum entropy $f(x)$ and the sample pdf, $f_s(x)$, which is derived from the data. To this end, the Kullback-Leibler measure of information variation is used :

Table 2.3

Information functions

x	$[\ln(x-a)]^2$
x^2	x^3
$\ln x$	x^4
$\ln(x-a)$	$\ln(x^2+1)$
$\ln(b-x)$	

$$I[f(x), f_s(x)] = \int_0^\infty f(x) \ln \left(\frac{f(x)}{f_s(x)}\right) dx. \tag{2.79}$$

Expression (2.79) measures the information loss or gain when the pdf $f(x)$ replaces the sample pdf $f_S(x)$. The information exchange between the maximum entropy pdf and the sample pdf is measured by the information divergence [21]:

$$S[f(x), f_S(x)] = I[f(x), f_S(x)] + I[f_S(x), f(x)]. \tag{2.80}$$

The pdf $f(x)$ is determined by adding the information functions $g_i(x)$ in (2.76) one at a time, until the divergence (2.80) becomes smaller than a preselected convergence criterion. Once the $f(x)$ is defined in this information function addition phase, the process is repeated in reverse to eliminate unnecessary and redundant functions. Let $\tilde{f}(x)$ be the maximum entropy pdf with one less information function. The divergence $S[f(x), \tilde{f}(x)]$ is calculated and the respective function is eliminated from the active set if the divergence is smaller than the criterion.

Finally, the coefficients β_i are calculated by solving the constraint equations (2.77) with the C_i values given by (2.78) and, thus, the distribution law is completely specified.

The last method described suggests that, instead of selecting a specific distribution law among the available set, a general law should be developed, which could be adapted to any system by an adequate choice of the parameters. A distribution law of such generality as to be able to describe any failure mechanism would contain a large number of parameters and the difficulty of modelling would consist of specifying their values. A general distribution law can at most approximate the true distribution of the time-of-failure, as it uses as few parameters as possible. The essential element in developing a general distribution is the exponential distribution. A mixture or succession of exponentials can approximate to any level of accuracy, the true distribution law of the time to failure.

A mixture of exponentials may be developed in several ways. For instance, any pdf may be written as a linear combination [22]:

$$f(t) = \sum_{i=1}^{n} \omega_i \lambda_i e^{-\lambda_i t} \tag{2.81}$$

where

$$\sum_{i=1}^{n} \omega_i = 1.$$

Relation (2.81) may be interpreted as follows. Consider a system with n mutually exclusive failure modes. Each failure mode is described by an exponential distribution of the time-to-failure with parameter λ_i, and ω_i is the probability that the system fails in mode i. Such a system has a pdf of the form (2.81). This is a model with stages in parallel.

Another mixture of exponentials appears when the system passes through several stages before it fails (stages in series). The time-to-failure is the sum of the times spent in the successive intermediate states. If each time is exponentially distributed with parameters λ_i, the Laplace transform of the time-of-failure distribution is:

$$f^*(s) = \mathscr{L} f(t) = \prod_{i=1}^{n} \frac{\lambda_i}{s + \lambda_i} . \tag{2.82}$$

The resulting pdf is a linear combination of exponentials if the parameters λ_i are distinct, or a linear combination of gamma distributions if some of the λ_i are equal. When all λ_i are equal, the pdf $f(t)$ is of the gamma type with parameters n and λ. The intermediate states in the model (2.82) can successfully describe the wearing-out of a system.

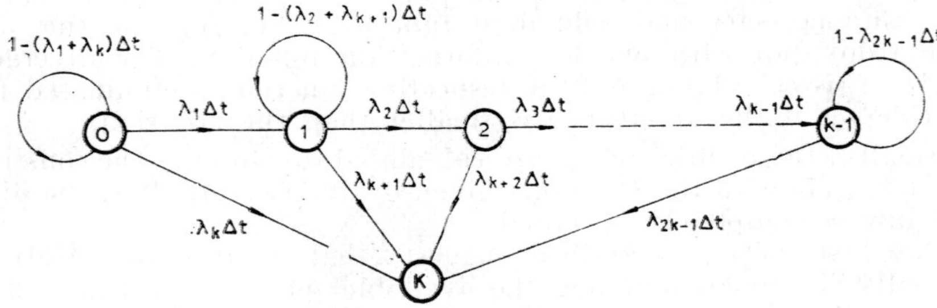

Fig. 2.19. — The method of stages.

A combination of series stages in parallel leads to a pdf having the Laplace transform:

$$f^*(s) = \sum_{i=1}^{n} \omega_i \left(\frac{\lambda_i}{s + \lambda_i} \right)^{k_i} . \tag{2.83}$$

The resulting pdf is a multi-modal function [24].

Let us consider a system with $(k - 1)$ intermediate states between the initial state, 0, and the failed state, k. If the failure state can be reached from any of the working states, we obtain the model depicted in Fig. 2.19, where the nodes represent the states and the branches the transition possibilities. A quantitative analysis of this model may be performed by the method of Markov chains (see Chapter 5). Denoting by $\lambda_1, \lambda_2, \ldots, \lambda_{k-1}$ the transition rates between the working states and by $\lambda_k, \lambda_{k+1}, \ldots \lambda_{2k-1}$ the transition rates to the failure state, the probability that the system is in state i at time t is [22]:

$$\begin{aligned} P_i(t) &= \sum_{j=1}^{i} D_{ij} \exp\left[-(\lambda_j + \lambda_{k+j})\, t\right] \\ D_{ij} &= \prod_{l=1}^{i-1} \lambda_l \left[\prod_{\substack{l=1 \\ l \neq j}}^{i} (\lambda_l - \lambda_j + \lambda_{k+l} - \lambda_{k+j}) \right]^{-1} . \\ & \qquad i = 0, 1, 2, \ldots, k-1. \end{aligned} \tag{2.84}$$

Summing (2.84) over i, we find the reliability function of the model.

The method of stages provides an exact representation of any pdf with a rational Laplace transform and accurate approximations for other pdfs. Expressions (2.81) to (2.84) use a small number of terms to approximate accurately the usual distribution laws. The parameters λ_i must be estimated from the data and a goodness-of-fit test is still necessary.

Another general distribution law can be obtained by discretely approximating the true failure rate by a piecewise constant function. The pdf of the time-to-failure is then a succession of exponential distributions. Such an approximation of a bath tub failure rate is shown in Fig. 2.20. It may be noted that this general distribution can model IFR, DFR and CFR systems as well.

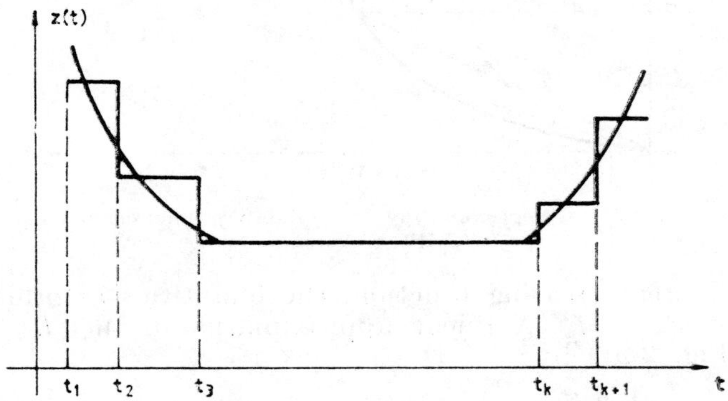

Fig. 2.20. — Piecewise exponential failure rate.

The reliability function of a system modelled by a succession of exponential distributions is :

$$R^*(t) = \begin{cases} \exp\left\{-\left[\sum_{j=1}^{i-1} \lambda_j(t_j - t_{j-1}) + \lambda_i(t - t_i)\right]\right\}; & t_{i-1} \leqslant t < t_i; \\ & i = 1, 2, \ldots, n \\ 0; \qquad t \geqslant t_n & t_0 = 0. \end{cases} \tag{2.85}$$

We shall prove in the sequel that any reliability function $R(t)$ can be approximated by a function $R^*(t)$ of the form given by (2.85), that is, for any t and ε, we may write :

$$|R^*(t) - R(t)| < \varepsilon. \tag{2.86}$$

Let us consider an IFRA system and its star-shaped $\ln(1/R)$ function, as depicted in Fig. 2.21. As $R(t)$ is decreasing there exists, for any $\varepsilon > 0$, a value t_ε such that for $t > t_\varepsilon$, $R(t) < \varepsilon$. It is obvious that for $t > \max(t_n, t_\varepsilon)$, $R(t) < \varepsilon$ and $R^*(t) = 0$, so that the inequality (2.86) is satisfied. Inequality (2.86) must now be proved for $t < t_n$. Let $\delta =$

$= \ln(1 + \varepsilon)$ and denote by $\{t_i ; \ i = 1, 2, \ldots, n\}$ the solution of :

$$\ln[1/R(t)] = i\delta ; \quad i = 1, 2, \ldots, n. \tag{2.87}$$

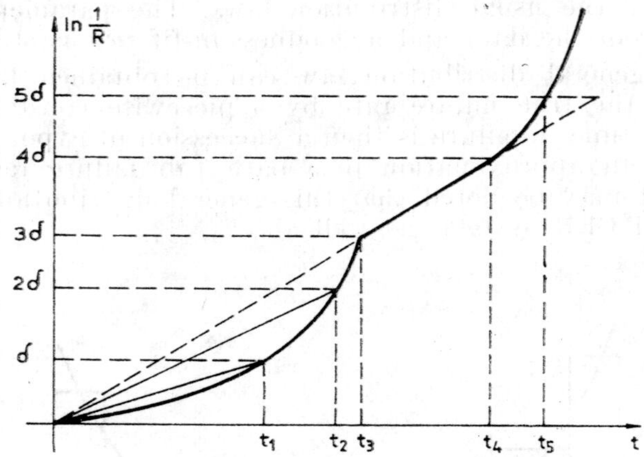

Fig. 2.21. — Approximating the reliability function of an IFRA system.

As $\ln(1/R)$ is an increasing function, the quantities t_i form an ordered set $t_1 < t_2 < \ldots < t_n$. A linear approximation of $\ln(1/R)$ in $(0, t_n)$ is given (see Fig. 2.21) by :

$$\ln(1/R^*) = \begin{cases} \dfrac{\delta}{t_1} t & 0 \leqslant t < t_1 \\ \dfrac{2\delta}{t_2} t & t_1 \leqslant t < t_2 \\ \cdot & \\ \cdot & \\ \dfrac{i\delta}{t_i} t & t_{i-1} \leqslant t < t_i \\ \cdot & \\ \cdot & \\ \dfrac{n\delta}{t_n} t & t_{n-1} \leqslant t < t_n \end{cases} \tag{2.88}$$

Comparing (2.85) and (2.88), it follows that :

$$\lambda_i = \frac{i\delta}{t_i} ; \quad i = 1, 2, \ldots, n. \tag{2.89}$$

Expressions (2.89) define the parameters of the general reliability function $R^*(t)$. As the function $\ln(1/R)$ is star-shaped, the values $i\delta/t_i$, and consequently λ_i, are increasing with i. In a time interval $[t_{i-1}, t_i)$, the functions $\ln(1/R)$ and $\ln(1/R^*)$ satisfy the inequalities:

$$(i-1)\delta \leqslant \ln(1/R) \leqslant i\delta;$$

$$\frac{i\delta t_{i-1}}{t_i} \leqslant \ln(1/R^*) \leqslant i\delta \qquad (2.90)$$

$$i = 1, 2, \ldots, n.$$

From (2.90) we get

$$\ln(1/R^*) - \ln(1/R) \leqslant \delta. \qquad (2.91)$$

On the other hand, the linear function $\ln(1/R^*)$ is greater than the star-shaped $\ln(1/R)$, so:

$$\ln(1/R^*) - \ln(1/R) \geqslant 0. \qquad (2.92)$$

From (2.91) and (2.92):

$$1 \leqslant \frac{R}{R^*} \leqslant e^{\delta} = 1 + \varepsilon \qquad (2.93)$$

$$0 \leqslant \frac{R - R^*}{R^*} \leqslant \varepsilon.$$

Inequalities (2.93) prove that, for any $\varepsilon > 0$, the distribution law of the time-to-failure can be approximated from below by a succession of exponential distributions, with an error less than ε. If ε is small, δ is small and the number of parameters of the model is large.

A DFRA system may be described by a similar model. The true reliability function is approximated from above, with any degree of accuracy.

The principle of approximating a distribution law by a succession of exponentials is the same as that of replacing a continuous function by a piecewise constant function. The approximation is better when more exponential distributions are used.

The piecewise exponential distribution may also be justified from the point of view of informational statistics [18, 19]. Let us assume that, in a reliability test, n times-to-failure $t_1 < t_2 < \ldots < t_n$ are observed. The test time is then partitioned into N intervals of length Δt, such that each interval contains at most one failure. Let $\alpha_1 = n/N$ be the fraction of the intervals which contain one failure and $\alpha_0 = 1 - \alpha_1$.

We now test the hypothesis that the system is CFR by an informational test. That is, we shall compute the coefficient of informational correlation between the theoretical distribution and the empirical one.

The probability that no failure occurs in a time interval Δt is $e^{-\lambda\Delta t}$, given that the system is CFR, and the probability of exactly one failure occurring in the time interval is $\lambda\Delta t e^{-\lambda\Delta t}$. The observed corresponding probabilities are α_0 and α_1. According to (2.65), the coefficient of informational correlation is:

$$R_{S,T} = \frac{\alpha_0 \, e^{-\lambda \Delta t} + \alpha_1 \lambda \Delta t \, e^{-\lambda \Delta t}}{\sqrt{\alpha_0^2 + \alpha_1^2} \sqrt{e^{-2\lambda \Delta t}(1 + \lambda^2 \Delta t^2)}} =$$

$$= \frac{\alpha_0 + \alpha_1 \lambda \Delta t}{\sqrt{\alpha_0^2 + \alpha_1^2} \sqrt{1 + \lambda^2 \Delta t^2}}. \tag{2.94}$$

The parameter λ of the hypothesized exponential distribution must have such a value as to make the distribution best fit the data. The value of λ is then selected according to the condition:

$$\frac{\partial R_{S,T}}{\partial \lambda} = 0$$

which gives:

$$\lambda = \frac{\alpha_1}{\alpha_0 \Delta t}. \tag{2.95}$$

The maximum value of $R_{S,T}$ follows from (2.94) and (2.95):

$$\max_{\lambda} R_{S,T} = 1. \tag{2.96}$$

The result (2.96) shows that any system is CFR over sufficiently small time intervals, that is any distribution may be approximated by a piece-wise exponential distribution with any degree of accuracy, provided the division into time intervals is sufficiently fine.

2.4 Specification of the global model

The global reliability model is completely specified if the parameters of the time-to-failure distribution are known. These parameters have to be estimated from life test data. The difficulty of the estimation increases with the number of parameters, so complete specification of a reliability model is more complicated when the model requires a large number of parameters to make it adequate for a variety of situations.

As the general distribution laws are mixtures and linear combinations of exponentials, we shall present the problem of parameter estimation with special reference to the exponential distribution, that is to CFR systems. Details concerning other distribution laws from Table 2.2 may be found in the literature [5, 10, 14, 25].

According to statistical terminology, a point estimator $\hat{\theta}$ of a parameter θ is a function of the sample data which provides a rough estimate of the parameter.

Among the methods of deriving point estimators, one of the best known is the maximum likelihood method. According to it, the point estimate is the value which maximizes the probability that the actual data should be obtained. This probability is the likelihood function

$L(X/\theta)$, where X is the data vector and θ the parameter to be estimated. Let us suppose, as an example of maximum likelihood estimation (MLE), that r failures were observed when a sample of size n was tested during time T. The probability of observing r failures is given by the binomial distribution [10]:

$$L(r/F) = C_n^r F^r (1 - F)^{n-r}. \tag{2.97}$$

$F = F(T)$ is a parameter of the binomial distribution; it represents the probability of failure in $(0, T)$, that is the cdf of the time-to-failure. The MLE of F is obtained by maximizing the logarithm of the likelihood function (2.97):

$$\frac{\partial \ln L(r/F)}{\partial F} = 0; \qquad \hat{F} = \frac{r}{n}. \tag{2.98}$$

The MLE of other reliability measures may be computed from expression (2.98), using the relationships from Table 2.1.

It is important to note that the value of the cdf for $t = T$ was estimated without any concern for the type of distribution law. Such methods are called non-parametric, because the quantity of interest is a parameter of the likelihood function, not of the time-to-failure distribution. The disadvantage of non-parametric estimation is that no extrapolation is possible outside the time interval of the test.

We shall now develop an MLE estimator of the failure rate of a CFR system. This is a parametric estimation, because the time-to-failure is assumed to be exponentially distributed. The likelihood function depends in this case on the way the data are censored, that is on the plan of the experiment.

There are two types of censoring: type I, when the duration of the experiment is fixed and type II, when the number of failures is fixed. In both cases, the failed elements may or may not be replaced by new ones. This classification leads to four ways of organizing an experiment. Detailed monographs [10] describe parameter estimation in all these four cases. We shall restrict ourselves to type II censored samples without replacement and we shall extend the results to other kinds of test plans.

The data vector includes the r observed times to failure $(t_1, t_2, \ldots, t_r)$. The likelihood function is obtained by multiplying the pdfs of the times to failure and the survival probabilities of the non-failed elements. Taking into account also the ordering of the observed failures, we get:

$$L(t_1, t_2 \ldots t_r/\lambda) = A_n^r(\lambda e^{-\lambda t_1})(\lambda e^{-\lambda t_2}) \ldots (\lambda e^{-\lambda t_r})(e^{-\lambda t_r})^{n-r} = $$
$$= A_n^r \lambda^r e^{-\lambda T_\Sigma}; \tag{2.99}$$

where

$$T_\Sigma = \sum_{i=1}^{r} t_i + (n - r)\, t_r\,.$$

The total time on test T_Σ is a random variable which contains a synthesis of all the experimental information, which is called sufficient statistics for λ. The logarithm of the likelihood function may be written as a function only of the total time on test :

$$\ln L(T_\Sigma/\lambda) = \ln A_n^r + r \ln \lambda - \lambda T_\Sigma. \tag{2.100}$$

The MLE estimator of λ is obtained by maximizing (2.100) :

$$\frac{\partial \ln L(T_\Sigma/\lambda)}{\partial \lambda} = 0\,;$$

$$\hat{\lambda} = \frac{r}{T_\Sigma}\,. \tag{2.101}$$

Some desirable properties of the point estimator may be achieved by using an appropriate estimation method. Let $\hat{\theta}$ be an estimator of θ and $g(\hat{\theta}/\theta)$ its pdf. The estimator is unbiased if its expected value is equal to the actual value of the parameter :

$$\int_0^\infty \hat{\theta}\, g(\hat{\theta}/\theta)\, d\hat{\theta} = \theta. \tag{2.102}$$

The MLE estimator of F (2.98) is unbiased because the expected value of the binomially distributed number of failures is exactly nF.

In order to establish if the MLE estimator (2.101) is biased or not, it is necessary to derive the distribution law of the total time on test, T_Σ. Let T_Σ be written in the form :

$$T_\Sigma = \sum_{i=1}^{r} t_i + (n - r)\, t_r = n t_1 + (n - 1)(t_2 - t_1) +$$

$$+ (n - 2)(t_3 - t_2) + \ldots (n - r + 1)(t_r - t_{r-1}) = \sum_{i=1}^{r} s_i \tag{2.103}$$

where :

$$s_1 = n t_1$$

$$s_2 = (n - 1)(t_2 - t_1)$$

$$\ldots\ldots\ldots\ldots$$

$$s_r = (n - r + 1)(t_r - t_{r-1}).$$

The pdf of the random vector $(s_1, s_2, \ldots, s_r)$ is [12]:

$$f(s_1, s_2, \ldots, s_r/\lambda) = \frac{L(t_1, t_2, \ldots, t_r/\lambda)}{\begin{vmatrix} \frac{\partial s_1}{\partial t_1} & \frac{\partial s_1}{\partial t_2} & \cdots & \frac{\partial s_1}{\partial t_r} \\ \vdots & & & \\ \frac{\partial s_r}{\partial t_1} & \frac{\partial s_r}{\partial t_2} & \cdots & \frac{\partial s_r}{\partial t_r} \end{vmatrix}} =$$

$$= \frac{A_n^r \lambda^r e^{-\lambda T_\Sigma}}{\begin{vmatrix} n & 0 & 0 \ldots\ldots\ldots & 0 \\ -(n-1) & n-1 & \vdots & 0 \\ \vdots & \vdots & \vdots & \vdots \\ 0 & 0 & -(n-r+1) & (n-r+1) \end{vmatrix}} = \lambda^r e^{-\lambda \sum_{i=1}^{r} s_i}. \tag{2.104}$$

Expression (2.104) implies that $s_1, s_2, \ldots, s_r$ are iid (independent and identically distributed) random variables, exponentially distributed with the same parameter λ. The total time on test, being the sum of r iid exponential variables, is gamma distributed with parameters r and λ [26]:

$$g(T_\Sigma/\lambda) = \frac{(\lambda T_\Sigma)^{r-1} e^{-\lambda T_\Sigma}}{(r-1)!} \tag{2.105}$$

The expected value of $\hat{\lambda}$ may now be computed using equations (2.101) and (2.105)

$$E_{\hat{\lambda}} = \int_0^\infty \frac{r}{T_\Sigma} \frac{\lambda(\lambda T_\Sigma)^{r-1} e^{-\lambda T_\Sigma}}{(r-1)!} \, dT_\Sigma = \frac{r\lambda}{r-1}. \tag{2.106}$$

From (2.106), it can be seen that the MLE estimator of λ is biased. The bias is easily eliminated if equation (2.101) is modified as:

$$\hat{\lambda} = \frac{r-1}{T_\Sigma}. \tag{2.107}$$

The MLE estimators and their bias are similarly derived for other test plans. The results are given in Table 2.4.

Consistency and accuracy are other desirable properties of the point estimators. A point estimator is consistent if it tends to the actual value of the parameter as the sample size tends to infinity. MLE estimators generally are consistent [13]. The accuracy shows how close a point estimate is to the actual value of the parameter. A convenient measure of the accuracy is the loss function, defined as the expectation of the quadratic deviation of the estimator from the parameter to be estimated:

$$L_s(\theta) = E[(\hat{\theta} - \theta)^2] = \int_0^\infty (\hat{\theta} - \theta)^2 g(\hat{\theta}/\theta) \, d\hat{\theta}. \tag{2.108}$$

Table 2.4

MLE estimators of λ

Type of censoring	With replacement	Without replacement
I	$\dfrac{r}{T_{\Sigma}}$ (biased)	$\dfrac{r}{T_{\Sigma}}$ (biased)
	$T_{\Sigma} = nT.$	$T_{\Sigma} = \sum_{i=1}^{r} t_i + (n-r)T_{\Sigma}$
II	$\dfrac{r-1}{T_{\Sigma}}$ (unbiased)	$\dfrac{r-1}{T_{\Sigma}}$ (unbiased)
	$T_{\Sigma} = nt_r$	$T_{\Sigma} = \sum_{i=1}^{r} t_i + (n-r)\,t_r$

For unbiased point estimators, the quadratic loss function (2.108) is precisely the variance of the estimator. The variance is thus a measure of the accuracy. An unbiased estimator of minimum variance is called UMVUE (uniformly minimum variance unbiased estimator). An estimator $\hat{\theta}_0$ of θ is UMVUE if $E_{\hat{\theta}_0} = \theta$ and, for any θ, $V_{\hat{\theta}_0} < V_{\hat{\theta}}$, where $\hat{\theta}$ is any other unbiased estimator. The pdfs of three estimators of θ are depicted in Fig. 2.22. It can be seen that the estimator $\hat{\theta}_1$ is biased, while $\hat{\theta}_2$ and $\hat{\theta}_3$ are unbiased. The best estimator is $\hat{\theta}_3$, because it has the smallest variance.

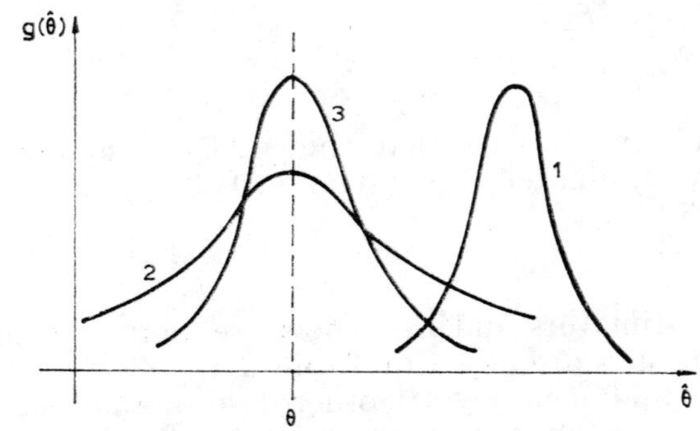

Fig. 2.22. — Properties of point estimators.

In the case of type II censored samples, the variance of the MLE estimator is:

$$V_{\hat{\lambda}} = \int_0^{\infty}\left(\frac{r-1}{T_{\Sigma}} - \lambda\right)^2 \frac{\lambda(\lambda T_{\Sigma})^{r-1}\mathrm{e}^{-\lambda T_{\Sigma}}}{(r-1)!}\,\mathrm{d}T_{\Sigma} = \frac{\lambda^2}{r-2}\,;$$

$$r > 2. \tag{2.109}$$

The accuracy increases with the number of observed failures. A related measure of the accuracy is the skewness of the estimator, defined as the ratio of the standard deviation to the actual value of the parameter:

$$\frac{\sigma_{\hat{\lambda}}}{\lambda} = \frac{1}{\sqrt{r-2}}; \quad r > 2. \tag{2.110}$$

The accuracy of the estimation is directly related to the quantity of information provided by the experiment. As the information increases, more is known about the actual value of the distribution parameter and the accuracy of the estimator improves. A measure of the quantity of information is thus also a measure of the estimation accuracy.

The quantity of information may be defined by starting from the distribution of the estimator or from the likelihood function.

According to Shannon [27], the quantity of information is defined as the expectation of the logarithm of the estimator distribution:

$$I_S = -\int_0^\infty g(\hat{\theta}) \log g(\hat{\theta})\, d\hat{\theta}. \tag{2.111}$$

Starting also from $g(\hat{\theta})$, Onicescu [18, 19] defined the informational energy as the integral of the square of the estimator distribution:

$$I_0 = \int_0^\infty g^2(\hat{\theta})\, d\hat{\theta}. \tag{2.112}$$

The informational energy (2.112) serves as a measure for the quantity of information, analogous to (2.111).

Shannon's measure of information and the informational energy have greater values when the distribution of the estimator is less uniform, so that the variance of the estimator is smaller and the parameter can be better located.

The information contained in an experiment may also be expressed by Fisher's definition, using the likelihood function [28]. When the experiment is highly informative, the likelihood function $L(X/\theta)$ must be very sensitive to slight changes in the value of the parameter θ, so that a clear distinction may be made between close values of θ. The quantity of information may then be expressed by the slope of the likelihood function with respect to θ. For ease of computation, the derivative of the logarithm of the likelihood function is considered. The measure of information must not depend on the data, so that it is necessary to take the mean value of the slope over the possible realizations of the data vector. As the slope can change its sign, the expected value of the square of the slope must be considered. Thus the definition of the information, in Fisher's sense, is the expectation of the square of the first derivative of the log-likelihood function with respect to the parameter θ :

$$I_F = \int_0^\infty \left[\frac{\partial \ln L(X/\theta)}{\partial \theta} \right]^2 f(X/\theta)\, dX. \tag{2.113}$$

We shall compute the informational energy and Fisher's information in the case of a type II censored sample. Similar calculations may be done for other test plans. From equations (2.105) and (2.112) we get the informational energy :

$$I_0 = \int_0^\infty \left[\frac{\lambda(\lambda T_\Sigma)^{r-1} e^{-\lambda T_\Sigma}}{(r-1)!} \right]^2 dt = \frac{\lambda^{2r}}{(r-1)!} \int_0^\infty T_\Sigma^{2r-2} e^{-2\lambda T_\Sigma}\, dT =$$

$$= \frac{\lambda^{2r}}{(r-1)!} \frac{\Gamma(2r-1)}{(2\lambda)^{2r-1}} = \frac{\lambda}{2^{2r-1}} \frac{r(r+1)\ldots(2r-2)}{(r-1)!}. \tag{2.114}$$

In order to compute Fisher's information, the likelihood function (2.99) and the distribution of the total time on test (2.105) are introduced into expression (2.113) :

$$I_F = \int_0^\infty \left[\frac{\partial \ln(A_n^r \lambda^r e^{-\lambda T_\Sigma})}{\partial \lambda} \right]^2 \cdot \frac{\lambda(\lambda T_\Sigma)^{r-1} e^{-\lambda T_\Sigma}}{(r-1)!}\, dT = \frac{r}{\lambda^2}. \tag{2.115}$$

It is interesting to note that Fisher's information is equal to the variance of the total time on test and approximately equal, for larger r, to the inverse of the variance of $\hat{\lambda}$. All measures of information show that the quantity of information increases with the number of failures.

The mean values like the variance or the different measures of information are not useful in the design of reliability tests for a prescribed accuracy level. To this end, another definition of accuracy is necessary, which should be more suitable for practical interpretations. Indeed, to prescribe a specific level for the variance or Fisher's information could be quite an arbitrary procedure because of the difficulty of using expected values to visualize the possible deviations of the estimate from the actual parameter.

According to [29], the accuracy of a point estimator must be defined on the basis of the normalized error $|\hat{\theta} - \theta|/\theta$, which is a random variable. Its distribution can be derived from $g(\hat{\theta})$. The accuracy is described in terms of the values the normalized error exceeds with prescribed small probabilities:

$$P\left(\frac{\hat{\theta}-\theta}{\theta}>q_1\right)=\varepsilon_1;\quad P\left(\frac{\theta-\hat{\theta}}{\theta}>q_2\right)=\varepsilon_2. \tag{2.116}$$

Relation (2.116) may also be written as:

$$P(\hat{\theta}>(1+q_1)\theta)=\varepsilon_1;\quad P(\hat{\theta}<(1-q_2)\theta)=\varepsilon_2. \tag{2.117}$$

From (2.117) we notice that $(1+q_1)\theta$ is the $(1-\varepsilon_1)$-percentile of the $g(\hat{\theta})$ distribution and $(1-q_2)\theta$ is the ε_2-percentile of the same distribution. ε_1 is the probability that the point estimate should exceed the actual value of θ by more than a fraction q_1 of this value. Similarly, ε_2 is the probability that the point estimate should fall short of the actual value of θ by more than a fraction q_2 of this value. The couples (q_1, ε_1) and (q_2, ε_2) define the accuracy on the right and on the left, respectively, as seen in Fig. 2.23.

This definition of estimation accuracy is consistent with the definitions in terms of mean values, discussed above. When the number of failures is small, there is a small amount of experimental information. The variance of the estimator is great and, consequently, the values of q_1 and q_2 are high, which means a low accuracy. The advantage of using definition (2.116) is that the values $(q_1, \varepsilon_1, q_2, \varepsilon_2)$ may be intuitively specified prior to the life test, and the number of failures may be computed accordingly.

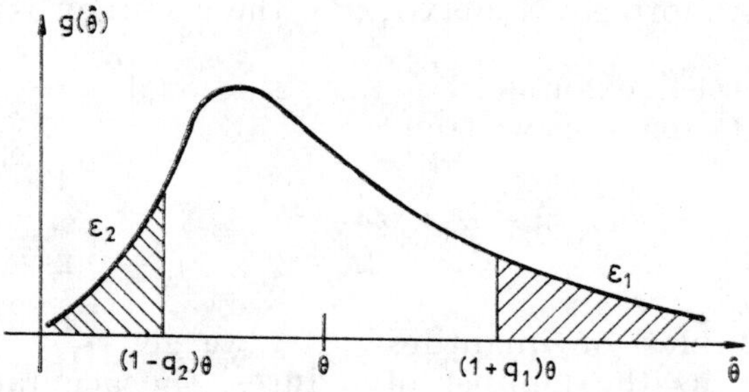

Fig. 2.23. — The accuracy of the point estimator.

We shall use definition (2.116) in the case of a type II censored sample in order to compute the number of failures necessary to meet a prescribed accuracy requirement. Using the point estimator (2.107) in (2.116) we get, after minor modifications:

$$P\left(T_\Sigma<\frac{r-1}{(1+q_1)\,\lambda}\right)=\varepsilon_1;\quad P\left(T_\Sigma>\frac{r-1}{(1-q_2)\,\lambda}\right)=\varepsilon_2. \tag{2.118}$$

Equations (2.118) may not be solved because the percentiles of $g(T_\Sigma/\lambda)$ depend on the unknown parameter. Let a new variable U be equal to $2\lambda T_\Sigma$. The pdf of U is:

$$f(u) = \frac{g(T_\Sigma/\lambda)}{\left|\frac{\mathrm{d}u}{\mathrm{d}T_\Sigma}\right|} = \frac{\lambda(\lambda T_\Sigma)^{r-1}e^{-\lambda T_\Sigma}}{2\lambda(r-1)!} = \frac{u^{r-1}e^{-\frac{u}{2}}}{2^r(r-1)!}. \tag{2.119}$$

The distribution (2.119) of $2\lambda T_\Sigma$ does not depend on λ. It is the well known χ^2 distribution with $2r$ degrees of freedom [12]. Using the variable $U = 2\lambda T_\Sigma$, relations (2.118) become:

$$P\left(U < \frac{2(r-1)}{1+q_1}\right) = \varepsilon_1; \qquad P\left(U > \frac{2(r-1)}{1-q_2}\right) = \varepsilon_2. \tag{2.120}$$

Relations (2.120) define two percentiles of the χ^2 distribution. If we denote by $\chi^2_\alpha(m)$ the α-percentile of a χ^2 distribution with m degrees of freedom, we have:

$$\frac{2(r-1)}{1+q_1} = \chi^2_{\varepsilon_1}(2r); \qquad \frac{2(r-1)}{1-q_2} = \chi^2_{1-\varepsilon_2}(2r). \tag{2.121}$$

From equations (2.121) we may compute the number of failures r for a prescribed accuracy level or, for a given r, we may compute the marginal errors q_1 and q_2 which are exceeded only with the prescribed probabilities ε_1, ε_2.

When the accuracy level is specified, equations (2.121) give different solutions r_1, r_2. Adopting $r = \max(r_1, r_2)$, the accuracy is better than the specified level.

As a numerical example, let $\varepsilon_1 = \varepsilon_2 = 5\%$, $q_1 = 30\%$, $q_2 = 20\%$. Equations (2.121) may be written:

$$\frac{\chi^2_{0.05}(2r)}{2r-2} = \frac{1}{1+0.3} = 0.769; \quad \frac{\chi^2_{0.95}(2r)}{2r-2} = \frac{1}{1-0.2} = 1.25. \tag{2.122}$$

Using the χ^2 tables of quantiles [31], we get $r_1 = 40$ and $r_2 = 50$. Adopting $r = 50$ as the number of failures, the accuracy on the right will be:

$$q_1 = \frac{98}{\chi^2_{0.05}(100)} - 1 = 25\% < 30\%. \tag{2.123}$$

Once the number of failures is adopted, the sample size and the test conditions should be arranged so that the failures can be accumulated in a reasonable time span.

In the case of type I censored samples, the duration of the test must be such that enough failures occur to assure the accuracy level.

The same equations (2.121) may be used to compute the accuracy but only prior information on the actual failure rate can give an estimate of the test time required.

It is often desirable to point out the uncertainty associated with the estimation of the parameters of the reliability model. The interval estimation provides a confidence interval (θ_L, θ_U) which contains the actual value of the parameter with a prescribed probability $1 - \gamma$, which is called the confidence level :

$$P(\theta_L < \theta < \theta_U) = 1 - \gamma. \tag{2.124}$$

The bounds θ_L, θ_U of the confidence interval are functions of the experimental data and are, consequently, random values. The confidence level is the probability that the random interval (θ_L, θ_U) covers the actual value θ of the parameter. In a sequence of samples from the same population, when a large number of confidence intervals are computed, only a fraction $1 - \gamma$ from this number contains the true parameter value.

The confidence interval as defined by (2.124) is two-sided. One-sided confidence intervals are also of interest. They are defined by the relations :

$$P(\theta_L < \theta) = 1 - \gamma; \quad P(\theta < \theta_U) = 1 - \gamma. \tag{2.125}$$

In principle, a confidence interval may be computed according to the following method [32]. Let $\hat{\theta}$ be a point estimation of θ and $g(\hat{\theta}/\theta)$ its pdf. When the pdf $g(\hat{\theta}/\theta)$ is known, two quantiles of $\hat{\theta}$ may be determined, such that :

$$P(\hat{\theta}_L < \hat{\theta} < \hat{\theta}_U) = 1 - \gamma. \tag{2.126}$$

As a rule, $\hat{\theta}_L$ and $\hat{\theta}_U$ are the $\gamma/2$ and $1 - \gamma/2$ percentiles of $\hat{\theta}$, respectively. They are represented in Fig. 2.24.

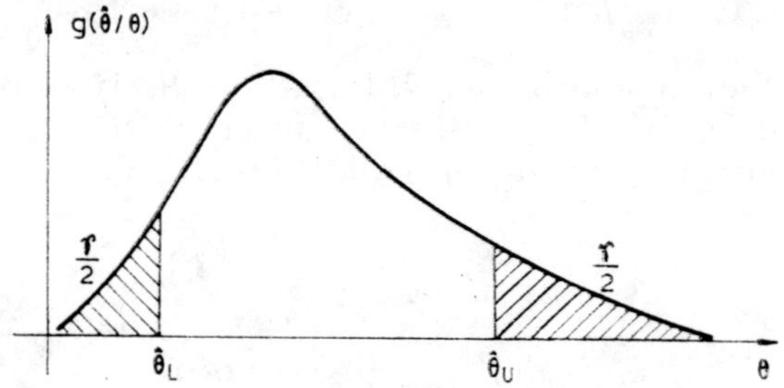

Fig. 2.24. — Point estimator distribution.

As the pdf of the point estimator depends on the actual value of the parameter θ, the quantiles $\hat{\theta}_L$ and $\hat{\theta}_U$ are increasing functions of θ :

$$\hat{\theta}_L = \hat{\theta}_L(\theta); \quad \hat{\theta}_U = \hat{\theta}_U(\theta). \tag{2.127}$$

These functions are depicted in Fig. 2.25.

For a fixed value θ_0 of the parameter, we have

$$P[\hat{\theta}_L(\theta_0) < \hat{\theta} < \hat{\theta}_U(\theta_0)] = 1 - \gamma. \tag{2.128}$$

Conversely, for a particular point estimate $\hat{\theta}$, two values θ_L and θ_U may be determined by inverting the functions (2.127), such that:

$$\hat{\theta} = \hat{\theta}_U(\theta_L) = \hat{\theta}_L(\theta_U). \tag{2.129}$$

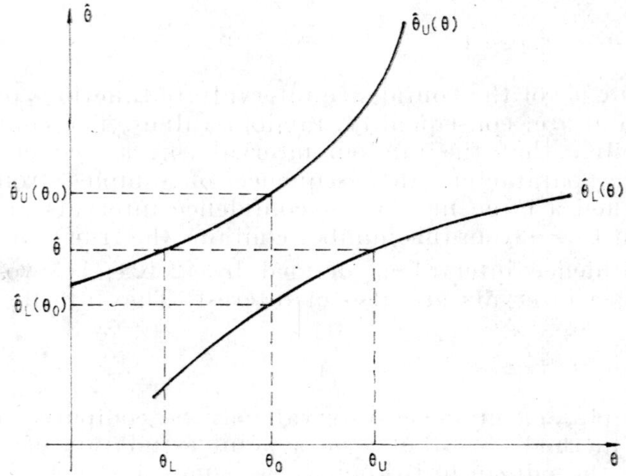

Fig. 2.25 — Confidence interval construction.

The interval (θ_L, θ_U) contains the value θ_0 of the parameter with the same probability that the point estimate $\hat{\theta}$ belongs to the interval $[\hat{\theta}_L(\theta_0), \hat{\theta}_U(\theta_0)]$.

$$P(\theta_1 < \theta_0 < \theta_U) = P[\hat{\theta}_L(\theta_0) < \hat{\theta} < \hat{\theta}_U(\theta_0)] = 1 - \gamma. \tag{2.130}$$

Thus, (2.130) defines a confidence interval of θ. In order to compute the bound θ_L and θ_U of the confidence interval for θ_0, the quantiles of $\hat{\theta}$ are first determined from the equations:

$$\int_0^{\hat{\theta}_L} g(\hat{\theta}/\theta)\, d\hat{\theta} = \gamma/2\, ; \quad \int_{\hat{\theta}_U}^{\infty} g(\hat{\theta}/\theta)\, d\hat{\theta} = \gamma/2. \tag{2.131}$$

From relations (2.129) and (2.131) we get the equations for θ_L and θ_U:

$$\int_0^{\hat{\theta}} g(\hat{\theta}/\theta_U)\, d\hat{\theta} = \gamma/2\, ; \quad \int_{\hat{\theta}}^{\infty} g(\hat{\theta}/\theta_L)\, d\hat{\theta} = \gamma/2. \tag{2.132}$$

We shall use equations (2.132) to derive the confidence interval for the probability of failure $F(T)$, in the context of the non-parametric estimation.

A point estimate of $F(T)$ is r/n, the fraction of failed elements. As the number of failures is binomially distributed, we have :

$$g(\hat{\theta}/\theta) = g(r/F) = C_n^r F^r (1-F)^{n-r}. \qquad (2.133)$$

From expressions (2.132) and (2.133), we get the equations of the two-sided confidence interval :

$$\sum_{k=0}^{r} C_n^k F_U^k (1-F_U)^{n-k} = \gamma/2\,; \quad \sum_{k=r}^{n} C_n^k F_L^k (1-F_L)^{n-k} = \gamma/2. \qquad (2.134)$$

A one-sided confidence interval demonstrates, with probability $1-\gamma$, that $F(T)$ is less than the upper value F_U :

$$\sum_{k=0}^{r} C_n^k F_U^k (1-F_U)^{n-k} = \gamma. \qquad (2.135)$$

In the parametric approach, a type II censored sample of CFR elements is considered. The point estimator of λ is r/T_Σ, and the pdf of T_Σ is of the gamma type, given by (2.105). The equations of the two-sided confidence interval of λ are then :

$$\int_0^{T_\Sigma} \frac{\lambda_L^r T_\Sigma^{r-1} e^{-\lambda_L T_\Sigma}}{(r-1)!}\, dT_\Sigma = \gamma/2\,; \quad \int_{T_\Sigma}^{\infty} \frac{\lambda_U^r T_\Sigma^{r-1} e^{-\lambda_U T_\Sigma}}{(r-1)!}\, dT_\Sigma = \gamma/2. \qquad (2.136)$$

Solving equations (2.136) would require extensive tabulation of the incomplete Euler function. Changing the variable T_Σ to $u = 2\lambda T_\Sigma$ transforms the gamma distribution into a χ^2 distribution with $2r$ degrees of freedom and equations (2.136) become :

$$\int_0^{2\lambda_L T_\Sigma} \frac{u^{r-1} e^{-u/2}}{2^r (r-1)!}\, du = \gamma/2\,; \quad \int_{2\lambda_U T_\Sigma}^{\infty} \frac{u^{r-1} e^{-u/2}}{2^r (r-1)!}\, du = \gamma/2. \qquad (2.137)$$

Using the quantiles of the χ^2 distribution, we get :

$$\lambda_L = \frac{\chi^2_{\gamma/2}(2r)}{2T_\Sigma}\,; \quad \lambda_U = \frac{\chi^2_{1-\gamma/2}(2r)}{2T_\Sigma}. \qquad (2.138)$$

Two one-sided confidence intervals of λ may be similarly derived :

$$\lambda_L = \frac{\chi^2_{\gamma}(2r)}{2T_\Sigma}\,; \quad \lambda_U = \frac{\chi^2_{1-\gamma}(2r)}{2T_\Sigma}. \qquad (2.139)$$

Formulae (2.138) and (2.139) still hold in the case of type II censored samples with replacement, but T_Σ must be computed accordingly. A slight modification is necessary in order to use them for type I censored samples.

Let us assume that, during the prescribed test time, r failures occurred. The test time of the type I censored sample is then limited by the durations of two type II censored samples at r and $r + 1$ failures respectively. We take the lower limit of the r-censored sample as the lower limit of λ and the upper limit of the $r + 1$-censored sample as the upper limit of λ. Thus the confidence interval covers the actual value of λ with a probability which is at least $1 - \gamma$:

$$\lambda_L = \frac{\chi^2_{\gamma/2}(2r)}{2T_\Sigma}; \quad \lambda_U = \frac{\chi^2_{1-\gamma/2}(2r+2)}{2T_\Sigma}. \tag{2.140}$$

The one-sided confidence intervals are:

$$\lambda_L = \frac{\chi^2_{\gamma}(2r)}{2T_\Sigma}; \quad \lambda_U = \frac{\chi^2_{1-\gamma}(2r+2)}{2T_\Sigma}. \tag{2.141}$$

The accuracy of interval estimation is directly connected with the length of the confidence interval. The larger the confidence interval, the smaller the accuracy of the estimation. On the other hand, the confidence interval is larger when the confidence level is high. For a given number of failures, a trade-off should be made between the confidence level and the accuracy of the estimation. As the number of failures increases, the confidence level and the accuracy also increase.

The accuracy may be quantitatively described using the distribution of the errors:

$$P\left(\frac{\theta_U - \theta}{\theta} > q_1\right) = \varepsilon_1; \quad P\left(\frac{\theta - \theta_L}{\theta} > q_2\right) = \varepsilon_2. \tag{2.142}$$

The two couples (q_1, ε_1) and (q_2, ε_2) represent the marginal errors. The probability that the upper limit of the confidence interval exceeds the actual value of the parameter by more than a fraction q_1 of its value is $1 - \varepsilon_1$. The probability that the lower limit is less than the actual value of the parameter by more than a fraction q_2 of its value is ε_2. Adopting convenient values for q_1, q_2, ε_1 and ε_2, the number of failures required may be computed from equations (2.142). We shall apply these equations for the confidence interval (2.140). In this case, equations (2.142) become:

$$P\left(2\lambda T_\Sigma < \frac{\chi^2_{1-\gamma/2}(2r)}{1+q_1}\right) = \varepsilon_1; \quad P\left(2\lambda T_\Sigma > \frac{\chi^2_{\gamma/2}(2r)}{1-q_2}\right) = \varepsilon_2. \tag{2.143}$$

As the random variable $2\lambda T_\Sigma$ is χ^2 distributed with $2r$ degrees of freedom, equations (2.143) define the ε_1 and the $1 - \varepsilon_2$ percentiles of the χ^2 distribution, respectively:

$$\chi^2_{\varepsilon_1}(2r) = \frac{\chi^2_{1-\gamma/2}(2r)}{1+q_1}\,; \quad \chi^2_{1-\varepsilon_2}(2r) = \frac{\chi^2_{\gamma/2}(2r)}{1-q_2}. \tag{2.144}$$

Equations (2.144) have different solutions r_1, r_2. Taking $r = \max\,(r_1, r_2)$ as a termination rule, the prescribed accuracy requirements are exceeded.

The accuracy in the case of the type I censored sample is described by the same equations (2.144). Their solution represents the number of failures which should be accumulated during the test time. Once this number is known, the total time on test and, hence, the duration and the sample size, may be approximated from equations (2.141) or (2.142) with some knowledge of the possible λ values.

As a numerical example, let $1-\gamma = 60\%$, $q_1 = 30\%$, $q_2 = 20\%$, $\varepsilon_1 = \varepsilon_2 = 10\%$. Equations (2.144) become:

$$\frac{\chi^2_{0.8}(2r)}{\chi^2_{0.10}(2r)} = 1.3\,; \quad \frac{\chi^2_{0.2}(2r)}{\chi^2_{0.9}(2r)} = 0.8. \tag{2.145}$$

The solution of (2.145) is $r = r_1 = r_2 = 50$. The life test will continue until 50 failures are accumulated, regardless of the test plan.

The above considerations of accuracy show that a small number of failures implies a low accuracy, even when the total time on test is large. The usual procedures to stop a life test at the first failure, or before any failure has occurred, neglect the accuracy aspect and consequently make it impossible to prove that a prescribed level of the failure rate has been achieved. Suppose that the true failure rate is λ_0 and a life test is performed in order to prove that λ is λ_0, with a confidence level $1-\gamma$. This means we have to compute the upper side of the confidence interval λ_U and verify that $\lambda_U < \lambda_0$. If the number of failures is low, the upper side of the confidence interval is far from $\lambda_0 (\lambda_U \gg \lambda_0)$, so the condition $\lambda_U < \lambda_0$ may not be verified even if $\lambda = \lambda_0$.

Often, instead of estimating the value of a parameter, one is interested in testing a hypothesis on the parameter value. It is tacitly admitted that hypothesis testing requires less experimental effort than parameter estimation, because in the experiment one seeks only grounds for rejecting or accepting a hypothesis with no concern about the more or less accurate localization of the true value of the parameter.

Let us consider, as a null hypothesis H_0, that the parameter θ is equal to θ_1, and as an alternative hypothesis H_1 that the value of θ is $\theta_2 > \theta_1$. The experimental data are required to distinguish between these two values of the parameter.

The decision rule is usually based on the point estimation $\hat{\theta}$, so that the null hypothesis is rejected if θ is greater than some decision criterion L_1 and accepted otherwise. The limit L_1 and the sample size are derived from the prescribed values of the risks, that is the probabilities of rejecting H_0, given that it is true, and of accepting H_0, given that it is false.

$$P(\hat{\theta} > L_1/\theta = \theta_1) = \alpha\,; \quad P(\hat{\theta} < L_1/\theta = \theta_2) = \beta. \tag{2.146}$$

Equations (2.146) may be written equivalently :

$$\int_{L_1}^{\infty} g(\hat{\theta}/\theta_1)\, d\hat{\theta} = \alpha ; \quad \int_{0}^{L_1} g(\hat{\theta}/\theta_2)\, d\hat{\theta} = \beta. \tag{2.147}$$

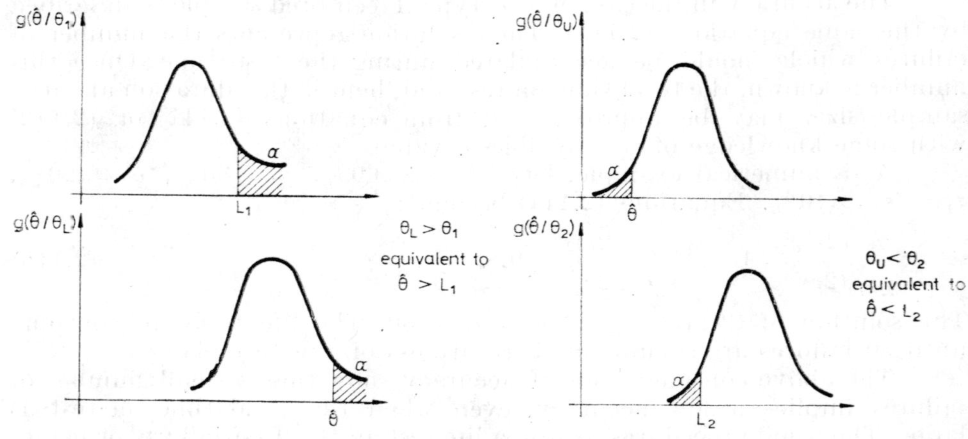

Fig. 2.26. — Interval estimation and hypothesis testing.

Comparing relations (2.147) and (2.132), and using Fig. 2.26 we may conclude that the rejection rule $\hat{\theta} > L_1$ is equivalent to $\theta_1 < \theta_L$, the type I risk α being equal to $\gamma/2$. The null hypothesis that $\theta = \theta_1$ is then rejected if the lower limit of the confidence interval is greater than θ_1. The probability of a type I error is half of the complement of the confidence level.

The type II risk β is related to the accuracy of the interval estimation. Let $\theta_1 = \theta_2(1 - q_2)$. The probability of accepting the null hypothesis $\theta = \theta_1$ when it is false is :

$$P[\theta_L < \theta_2(1 - q_2)/\theta = \theta_2] = \beta. \tag{2.148}$$

Comparing relation (2.148) to the second of relations (2.142), the risk β equals ε_2, the probability that the error should be greater than $q_2 = (\theta_2 - \theta_1)/\theta_2$.

Let us now take as a null hypothesis $\theta = \theta_2$ with the alternative $\theta = \theta_1 < \theta_2$. The null hypothesis is now rejected if the point estimate is greater than a limit L_2. The equations of the risks are :

$$P(\hat{\theta} < L_2/\theta = \theta_2) = \alpha ; \quad P(\hat{\theta} > L_2/\theta = \theta_1) = \beta \tag{2.149}$$

or, equivalently :

$$\int_{0}^{L_2} g(\hat{\theta}/\theta_2)\, d\hat{\theta} = \alpha ; \quad \int_{L_2}^{\infty} g(\hat{\theta}/\theta_1)\, d\hat{\theta} = \beta. \tag{2.150}$$

The comparison between (2.150) and (2.152) illustrated in Fig. 2.26 shows that the rejection rule $\hat{\theta} < L_2$ is equivalent to $\theta_U < \theta_2$, the risk α being equal to $\gamma/2$. The null hypothesis is then rejected if the upper limit of the confidence interval is less than the hypothesized value of θ. The confidence level is equal to the complement of twice the type I risk. The type II risk β is related to the accuracy of the interval estimation. Let $\theta_2 = \theta_1(1 + q_1)$. The probability of accepting the null hypothesis $\theta = \theta_2$ given that it is false is:

$$P[\theta_U > \theta_1(1 + q_1)/\theta = \theta_1] = \beta. \tag{2.151}$$

Comparing expression (2.151) with the first of the (2.142) relations, the risk β equals ε_1. The probability of accepting the hypothesis $\theta = \theta_2$, when actually $\theta = \theta_1$, is equal to the probability that the upper side error should be greater than

$$q_1 = (\theta_2 - \theta_1)/\theta_1.$$

Testing a hypothesis on the value of a parameter is equivalent to interval estimation for the parameter. The confidence level is analogous to the type I risk and the accuracy to the power of the test. In most cases a one-sided confidence interval is sufficient. Suppose one wants to test $\theta = \theta_1$ against $\theta = \theta_2 > \theta_1$ with the risk of errors α and β prescribed. From experimental data, a lower confidence limit θ_L is derived for a confidence level $1 - \alpha$ and a marginal error $q_2 = (\theta_2 - \theta_1)/\theta_1$, which may be exceeded only with probability β. The hypothesis is rejected when the hypothesized value θ_1 is less than the lower confidence limit.

Suppose now that the alternative to $\theta = \theta_1$ is $\theta = \theta_2 < \theta_1$. An upper confidence limit is derived fom the data, for a confidence level $1 - \alpha$ and a marginal error $q_1 = (\theta_1 - \theta_2)/\theta_2$, which may be exceeded only with probability β. The hypothesis is rejected if the hypothesized value θ_1 is greater than the upper confidence limit θ_U. We shall apply these considerations to the hypothesis testing of the parameter of the exponential distribution. Suppose the value $\lambda = \lambda_1$ must be tested against $\lambda = \lambda_2 > \lambda_1$, with the risks of error α and β, using a type II censored sample without replacement. We choose the confidence level equal to $1 - \alpha$ and derive the number of failures from the accuracy requirement:

$$P\left(\frac{\lambda - \lambda_L}{\lambda} > q_2\right) = \beta$$
$$q_2 = \frac{\lambda_2 - \lambda_1}{\lambda_2}; \quad \lambda_L = \frac{\chi^2_\alpha(2r)}{2T_\Sigma}. \tag{2.152}$$

Equation (2.152) may be written equivalently:

$$P\left(2\lambda T_\Sigma > \frac{\chi^2_\alpha(2r)}{1 - q_2}\right) = \beta. \tag{2.153}$$

The solution of equation (2.153) is obtained by using the quantiles of the χ^2 distribution:

$$\frac{\chi^2_{1-\beta}(2r)}{\chi^2_{\alpha}(2r)} = \frac{1}{1-q_2} = \frac{\lambda_2}{\lambda_1}. \tag{2.154}$$

Once the number of failures is determined, the test is performed and the lower confidence limit is computed. The hypothesis $\lambda = \lambda_1$ is rejected if $\lambda_L > \lambda_1$, that is:

$$T_\Sigma < \frac{\chi^2_{\alpha}(2r)}{2\lambda_1}. \tag{2.155}$$

Suppose now that the value $\lambda = \lambda_1$ must be tested against $\lambda = \lambda_2 < \lambda_1$ using the same risks and the same type of life test. The confidence limit $1 - \alpha$ is adopted and the accuracy requirement is

$$P\left(\frac{\lambda_U - \lambda}{\lambda} > q_1\right) = \beta, \tag{2.156}$$

where: $\lambda_U = \dfrac{\chi^2_{1-\alpha}(2r)}{2T_\Sigma}$; $q_1 = \dfrac{\lambda_2 - \lambda_1}{\lambda_1}$. Manipulation of equation (2.156) leads to:

$$P\left(2\lambda T_\Sigma < \frac{\chi^2_{1-\alpha}(2r)}{1+q_1}\right) = \beta. \tag{2.157}$$

Equation (2.157) defines the β percentile of the χ^2 distribution:

$$\frac{\chi^2_{1-\alpha}(2r)}{1+q_1} = \chi^2_{\beta}(2r). \tag{2.158}$$

The number of failures is, then, the solution of the equation:

$$\frac{\chi^2_{1-\alpha}(2r)}{\chi^2_{\beta}(2r)} = \frac{\lambda_2}{\lambda_1}. \tag{2.159}$$

The rejection rule $\lambda_1 > \lambda_U$ may be written:

$$T_\Sigma > \frac{\chi^2_{1-\alpha}(2r)}{2\lambda_1}. \tag{2.160}$$

It is interesting to note that in both cases, the number of failures is practically the same. (It is rigorously the same for $\alpha = \beta$.) If we want to prove that the failure rate is not worse than λ_1 (which means rejecting the alternative hypothesis that $\lambda = \lambda_2 > \lambda_1$), the total time on test must be greater than $\chi^2_{\alpha}(2r)/2\lambda_1$, while if we try to prove that the failure is better than λ_1 (which means accepting the alternative hypothesis that

$\lambda = \lambda_2 < \lambda_1$), the total time on test must be greater than $\chi^2_{1-\alpha}(2r)/2\lambda_1$, a condition more difficult to meet.

However, the experimental effort involved in testing hypotheses is by no means less than that implied by interval estimation, if the estimation accuracy is equivalent to the probabilities of statistical errors.

The problem of parameter estimation was described above with special reference to the case of CFR systems. As we have already seen, the exponential model attached to such systems plays an important role in reliability theory because any reliability model may be approximated by various combinations of exponential distributions.

We shall now mention some peculiarities of parameter estimation in the case of piecewise exponential distributions.

The MLE point estimator of the interval failure rate is :

$$\hat{\lambda}_i = \frac{r_i}{T_{\Sigma_i}}, \tag{2.161}$$

where r_1 and T_{Σ_i} correspond to the time interval $[t_{i-1}, t_i)$. In [33] it is shown that the random variables $\sqrt{n}(\hat{\lambda}_i - \lambda_i)$ are asymptotically independent, irrespective of the time-to-failure distribution. Moreover, their asymptotic distributions are normal with means equal to zero, and their variances dependent on the partition $\{t_i\}$. The specification of this partition is essential in the edification of the reliability model. The finer the partition, the more accurate the approximation of the time-to-failure distribution by the piecewise exponential function.

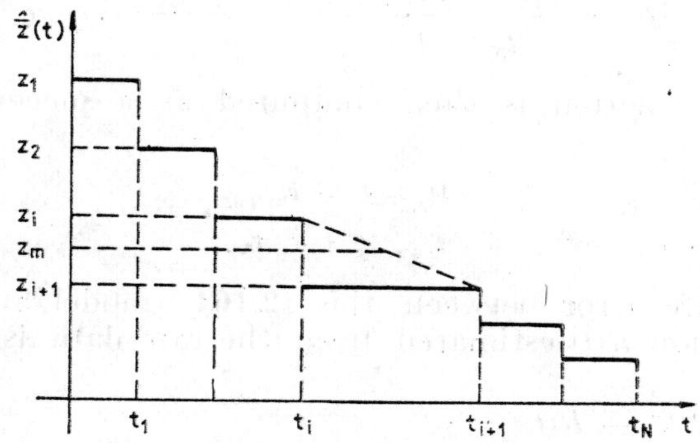

Fig. 2.27. — Empirical failure rate.

In [34], a method to determine the partition at a specified level of accuracy is presented. It uses a histogram of the data like the one depicted in Fig. 2.27. The intial partition in Fig. 2.27 is given by the times $t_1, t_2, \ldots, t_N$, when the elements of the sample are inspected and the failure rate in $[t_{i-1}, t_i)$ is :

$$\hat{z}(t_{i-1}, t_i) = \frac{r_i}{n(t_{i-1})(t_i - t_{i-1})} \tag{2.162}$$

where $n(t_{i-1})$ is the number of non-failed elements at t_{i-1}. The simplest way to specify a general distribution law of the type given by the expression (2.85) is to take the inspection times as discontinuities of the failure rate. The number of parameters could be too large with respect to the required accuracy of the approximation. The intervals determined by the final partition will be called pseudo-intervals, in order to distinguish them from the intervals between the inspection times.

In the case of IFR or DFR systems, at least two pseudo-intervals are required. The case of a bath tub failure rate is treated similarly considering its monotonic intervals.

The two pseudo-intervals from Fig. 2.27 which represent a DFR system are easily determined, taking as a discontinuity point the final point of the interval where the observed failure rate is closest to its median, z_m. This point is t_{i+1} in Fig. 2.27 and the two pseudo-intervals are $(0,\ t_{i+1})$ and $(t_{i+1},\ t_N)$. The number of pseudo-intervals may be increased, dividing each pseudo-interval by the same method. The number of pseudo-intervals is at most N, equal to the inspection times.

The parameters λ_i associated with the pseudo-intervals are estimated by mediating the point estimates (2.162). In the case illustrated in Fig. 2.27, the parameters of the two pseudo-intervals are :

$$\begin{aligned}\lambda_1 &= \frac{z_1 t_1 + z_2(t_2 - t_1) + \ldots + z_{i+1}(t_{i+1} - t_i)}{t_{i+1}} \\ \lambda_2 &= \frac{z_{i+2}(t_{i+2} - t_{i+1}) + \ldots + z_N(t_N - t_{N-1})}{t_N - t_{i+1}} .\end{aligned} \tag{2.163}$$

The reliability function is then computed as a piecewise exponential function :

$$R(t) = \begin{cases} e^{-\lambda_1 t} & 0 \leqslant t < t_{i+1} \\ e^{-[\lambda_1 t_{i+1} + \lambda_2 (t - t_{i+1})]} & t_{i+1} \leqslant t < t_N . \end{cases} \tag{2.164}$$

The mean square error between the (2.164) model and the empirical reliability function $\hat{R}(t)$ estimated from the raw data is then computed

$$E = \sum_{i=1}^{N} [R(t_i) - \hat{R}(t_i)]^2 \tag{2.165}$$

where : $\hat{R}(t_i) = \dfrac{n(t_i)}{n}$; $i = 1,2, \ldots, N$.

The number of pseudo-intervals is increased until the mean square error (2.165) becomes less than a prescribed value.

Specification of the reliability model by classical statistical methods implies a considerable experimental effort in order to attain a reasonable accuracy both in approximating the time-to-failure distribution and in estimating its parameters. Most of the methods described have been devised assuming plentiful experimental information. The non classical approach which will be described in the next Chapter is intended to make the most of the available information to derive adequate models when experimental data are scarce.

References

1. Barlow, R. E., Proschan, F., A Mathematical Theory of Reliability, John Wiley, New York, 1965.
2. Barlow, R. E., Proschan, F., Statistical Theory of Reliability and Life Testing. Probability Models, Holt, Rinehart and Winston, New York, 1975.
3. Muth, E. J., Memory as a Property of Probability Distribution, IEEE Trans. Reliab., R-29, 2 (1980) 160–165.
4. Barlow, R. E., Campo, R., Total time on test processes and applications to failure data analysis. Reliability and Fault Tree Analysis, SIAM, Philadelphia, 1975.
5. Mann, N., Schafer, R., Singpurwalla, N.D., Methods for Statistical Analysis of Reliability and Life Data, John Wiley, New York, 1974.
6. Singpurwalla, N.D., Statistical models for reliability, IEEE Trans. Reliab., R-30, 3 (1971) 35–42.
7. Koai, C.F., Relation of a physical process to the reliability of electronic components, IEEE Trans. Reliab., R-25, 3 (1976) 11–27.
8. Alexanian, I.T., Brodie, D.E., A method for estimating the reliability of ICs, IEEE Trans. Reliab., R-26, 5 (1977) 359–361.
9. Shooman, M. I., Reliability physics models, IEEE Trans. Reliab., R-17, 1 (1968) 14–20.
10. Gnedenko, B.V., Beleaev, I.K., Soloviev, A.D., Mathematical Methods in Reliability Theory (in Romanian), Ed. tehnică, Bucharest, 1968.
11. Vodă, V., New Statistical Models in Reliability (in Romanian), Ed. Academiei, Bucharest, 1980.
12. Trebici, V. (ed.), Encyclopaedia of Statistics (in Romanian), Ed. științifică și enciclopedică, Bucharest, 1985.
13. Wiener, Ul., Vodă, V., Isaic-Maniu, Al., Probability Charts in Engineering (in Romanian), Ed. tehnică, Bucharest, 1983.
14. Pollard, R., Rivoire, P., Fiabilité et Statistique Prévisionelle : La Méthode de Weibull, Eyrolles, Paris, 1969.
15. Terzian, R.C., Comments on K-S Tests, IEEE Trans. Reliab., R-31, 2(1982) 210.
16. Barr, D.R., Davidson, T., A Kolmogorov-Smirnov test for censored samples, Technometrics, 15, 4 (1973) 739–757.
17. Schwob, M., Peyrache, G., Traité de Fiabilité, Mason, Paris, 1969.
18. Onicescu, O., Ștefănescu, V., Elements of Informational Statistics with Applications (in Romanian), Ed. tehnică, Bucharest, 1979.
19. Ștefănescu, V., Applications of Informatioual Energy and Correlation (in Romanian), Ed. Academiei, Bucharest, 1979.
20. Evans, R.A., The principle of minimum information, IEEE Trans. Reliab., R-18, 3 (1969), 87–91.
21. Miller, J.E., Kulp, R.W., Orr, G.E., Adaptive probability distribution based upon maximum entropy, IEEE Trans. Reliab., R-33, 4 (1984) 353–357.
22. Singh, C., Billinton, R., Lee, S.Y., The method of stages in non Markov models, IEEE Trans. Reliab., R-26, 2 (1977) 135–137.
23. Teodorescu, D., Optimized Stochastic Models (in Romanian), Ed. Academiei, Bucharest, 1982.
24. Singh, C., On the Characteristics of Method-of-Stages, IEEE Trans. Reliab., R-38, 1 (1979) 91–92.
25. Sinha, S.K., Kale, B.K., Life Testing and Reliability Estimation, Wiley Eastern, New Delhi, 1979.
26. Gnedenko, B.V., The Theory of Probability, Mir, Moscow, 1969.
27. Shannon, C.E., Weaver, W., The Mathematical Theory of Communication, Habana, 1949.

28. Canfield, R., Teed, J.C., Selecting the prior distribution in Bayesian estimation, IEEE Trans. Reliab., R-26, 4 (1977) 283—285.
29. Cătuneanu, V.M., Mihalache, A., Design of Reliability Tests (in Romanian), EEA, 30, 1 (1982) 3—15.
30. Mihalache, A., Estimation Accuracy and Hypotheses Testing (in Romanian), Buletin IPB, Electronica, 46—47 (1985) 147—152.
31. Kozlov, V., Usakov, I., Reliability Handbook (in Russian), Sovetskoe Radio, Moscow, 1975.
32. Brownlee, K.A., Statistical Theory and Methodology in Science and Engineering, John Wiley, New York, 1960.
33. Barlow, R.E., Proschan, F., Asymptotic theory of total time on test processes with applications to life testing, Preprint, University of California, Berkeley, 1976.
34. Yen, Y.C., Sheng, C.L., A new reliability function, IEEE Trans. Reliab., R-23, 2 (1974) 118—122.
35. Lch, W., A new generalization of the class of NBU distributions, IEEE Trans. Reliab., R-33, 5 (1984) 419—422.

Chapter 3

Bayesian modelling of system reliability

3.1 Parameter estimation

The global modelling of system reliability, described in Chapter 2, led to the adoption of a generally applicable life distribution which is a combination of exponential distributions with different parameters. Complete specification of the model implies estimation of the parameters of these distributions. Reasonable precision in this estimation is attained only when a large amount of experimental data is available. In current applications, the sample and the duration of the reliability test are limited by economic considerations, so a method is needed to improve the precision of the estimation when experimental data are scarce.

The basic idea of this method is that, before performing any reliability test, a certain amount of information on the reliability of the system should be available. This piece of information, when used in combination with the experimental data, would improve the precision of the reliability model. The methods of parameter estimation which take into account the entire information available on the parameters (be it experimental or not) are called bayesian methods. They play an important role in modern statistics, on account of their suitability for situations when conclusions must be drawn although insufficient evidence is provided [2, 3]. The development of bayesian methods was closely connected with that of reliability theory, which provided fertile interpretations and suggestions for the mathematics of bayesian statistics.

In reliability theory we have to estimate the parameters of the life distribution when its nature is known, or estimate the reliability measures directly when the nature of the distribution law is ignored (non-parametric estimation). We shall denote by θ either a parameter of the life distribution or a certain reliability measure. The information available on θ before the experiment, called a-priori information, should be expressed in a quantitative form. This may be done by associating a probability distribution $f_a(\theta)$ with the θ values which reflects the degree of belief of the analyst concerning the actual value of θ. When θ is continuous in the range $(0,\infty)$ the probability distribution $f_a(\theta)$ is a continuous pdf (Fig. 3.1.). The shaded area $f_a(\theta)\,d\theta$ represents the degree of belief of the analyst that the actual value of θ lies in the interval $(\theta, \theta + d\theta)$.

The association of a probability distribution with an unknown but deterministic measure is not correct in the classical interpretation of probability as a fraction or as the limit of a frequency. The procedure is justified only if probability is considered from a subjective point of view, as the degree of belief of an idealized, purely rational person.

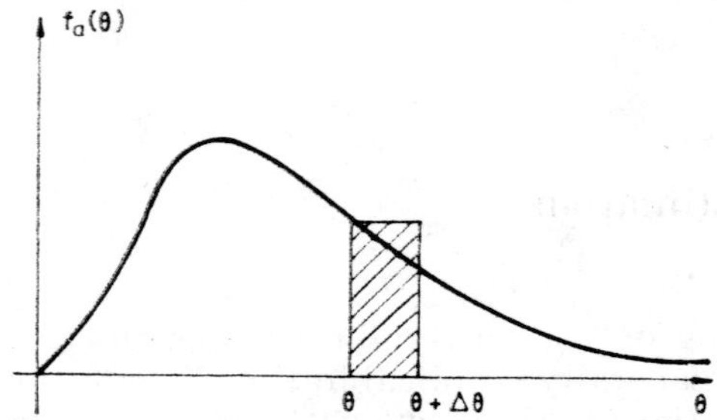

Fig. 3.1. — A-priori distribution.

It is necessary to specify the problem of estimation in decision theory when examining the validity of the subjective interpretation. Let us denote by $\delta(x)$ the decision about the actual value of θ, based on the experimental data x. To each pair $(\theta, \delta(x))$ we associate the loss $L(\theta, \delta(x))$ incurred by taking the decision δ when the actual value is θ. Before the experimental data are available, the decision is taken on the basis of the subjective probabilities associated with the θ values. In real decisions, the a-priori distribution is influenced by the subjectivity of the analyst, so that the decision-making is a matter of psychology. In our case, we are not interested in the behaviour of real people under uncertain circumstances, but in the way an idealized, completely rational person makes decisions only on the grounds of the objective information available to him. If we postulate the existence of such a behaviour, the objections against the subjective interpretation of probability lose strength as the a-priori distribution ceases to be arbitrary.

Carnap [1] has shown the characteristics an a-priori distribution should possess in order to be considered reasonable. First, an a-priori distribution must be strictly coherent, that is it must not lead to losses which would either be all positive or zero. This condition is fulfilled [1] if the subjective probabilities satisfy the axiom of Kolmogorov and if the a-priori distribution is annulated only for impossible values of θ. This condition is important because it shows that there are no formal differences between the objective and the subjective probabilities, both being different interpretations of the same mathematical function. It is essential that non-zero subjective probabilities should be assigned to all possible values of θ, otherwise the decision might be erroneous and the losses entirely positive.

Secondly, the a-priori distribution should have the property of symmetry, that is it should be unique under various kinds of information describing the same quality. Two different analysts with equal amounts of information must associate the same a-priori distribution with the values of θ. This condition is only partially satisfied at the present stage of bayesian estimation.

The third condition refers to the way the a-priori piece of information is combined with the experimental data. This combination leads to a transformation of the a-priori distribution, which should depend only on the data encountered. Modifications of the degree of belief of the analyst under the impact of the data must be accounted for only by his accumulated experience, not by any other subjective factors.

We shall assume in the following that the information initially available on θ may be expressed by an a-priori distribution $f_a(\theta)$ which satisfies Carnap's conditions of rationality. The entire range of parameter values is covered, and the uncertainty with respect to the actual value of θ is not artificially underrated. There are different measures of this uncertainty, such as the variance of the a-priori distribution, a confidence interval, and the amount of information in Shannon's or Fisher's sense. As the amount of data increases, the uncertainty decreases.

We shall describe the process of combining the a priori information with the experimental data. Let x be the vector of the times-to-failure and $f(x/\theta)$ its probability density function. At any stage of the process a decision $\delta(x)$ which depends on the data vector x may be taken concerning the θ value. The loss function $L(\theta, \delta(x))$ describes the accuracy of the decision. The general form of the loss function is :

$$L(\theta, \delta(x)) = \theta^{w_0}(\theta - \delta(x))^{w_1}. \tag{3.1}$$

The expected value of $L(\theta, \delta(x))$ with respect to the a-priori distribution and the various possible data is called the expected risk, $\rho(f_a, \delta)$, and may be written

$$\rho(f_a, \delta) = \int_{D_\theta}\int_{D_x} L(\theta, \delta(x)) f_a(\theta) f(x/\theta)\, dx d\theta \tag{3.2}$$

where D_θ and D_x are the ranges of θ and x respectively.

For a certain decision $\delta(x) = d$, the expected risk becomes :

$$\rho(f_a, d) = \int_{D_\theta} L(\theta, d) f_a(\theta)\, d\theta. \tag{3.3}$$

A decision function $\delta^*(x)$ is optimal (bayesian) with respect to a certain a-priori distribution if it minimizes the expected risk (3.2) :

$$\rho(f_a, \delta^*) = \inf_{\delta} \rho(f_a, \delta) = \rho^*(f_a). \tag{3.4}$$

Relation (3.4) shows that, for a certain amount of a-priori information, there is a unique decision which minimizes the risk and this minimal value depends on the information initially available.

The mathematical model of the process of combining the a-priori information with the experimental data is completed if the optimal decision $\delta^*(x)$ for any x is supplied. Let d^* be the optimal decision for a given set of experimental data, x. This decision minimizes the integral :

$$\rho(f_a, x) = \int_{D_\theta} L(\theta, d) f_a(\theta) f(x/\theta) \, d\theta. \tag{3.5}$$

On the other hand, the unconditional pdf of x is :

$$f_1(x) = \int_{D_\theta} f(x/\theta) f_a(\theta) \, d\theta. \tag{3.6}$$

As equation (3.6) does not depend on θ, the decision d^*, which minimizes expression (3.5), also minimizes

$$\frac{\rho(f_a, x)}{f_1(x)} = \int_{D_\theta} L(\theta, d) \frac{f_a(\theta) f(x/\theta)}{f_1(x)} \, d\theta. \tag{3.7}$$

From the Bayes equation, we can write :

$$f(\theta/x) = f_p(\theta) = \frac{f_a(\theta) f(x/\theta)}{f_1(x)}. \tag{3.8}$$

The expression (3.8) is the subjective distribution associated with θ after performing the experiment. It is called the a-posteriori distribution, and is the result of a synthesis between the a-priori information and the experimental data. From equations (3.8) and (3.7), we get

$$\rho(f_p, x) = \int_{D_\theta} L(\theta, \mathrm{d}) f(\theta/x) \, d\theta. \tag{3.9}$$

The expression (3.9) is the expected loss function conditioned by a given set of experimental data. For this set, the decision d^* which minimizes at the same time expressions (3.5) and (3.9) is a bayesian (optimal) estimation of θ with respect to a particular choice of the a-priori distribution. It is important to note that minimizing (3.9), yields particular values of $\delta^*(x)$ even when the nature of the function $\delta^*(x)$ is ignored. It is difficult to compute the optimal decision when the loss function is of the general form of relation (3.1). The most common loss function is of quadratic form, which results from function (3.1) by putting $w_0 = 0$, $w_1 = 2$:

$$\rho(f_p, x) = \int_{D_\theta} (\theta - \mathrm{d})^2 f(\theta/x) \, d\theta. \tag{3.10}$$

Minimizing formula (3.10) with respect to d, we get

$$d^* = \int_{D_\theta} \theta f(\theta/x)\, d\theta. \tag{3.11}$$

Thus, for a quadratic loss function, the bayesian estimation of θ is its expected posterior value, the minimal risk being, evidently, the posterior variance.

Another way of defining the loss function is

$$L(\theta, d) = |\theta - d|. \tag{3.12}$$

Minimizing the risk (3.9) for the loss function (3.12), we obtain the posterior median as the bayesian estimation of θ.

The optimal estimator is analogous to the point estimator, which was determined in Chapter 2 exclusively from experimental data. A confidence interval may also be defined by a relationship analogous to the classical definition :

$$P(\theta_L < \theta < \theta_U) = 1 - \gamma. \tag{3.13}$$

The interpretation of relationship (3.13) is essentially different from the classical case. Here, the confidence level $(1 - \gamma)$ is the subjective probability assigned to the interval (θ_L, θ_U). The confidence interval is computed from the a-posteriori distribution by the equations :

$$\int_0^{\theta_L} f_p(\theta)\, d\theta = \gamma/2 ; \quad \int_{\theta_U}^{\infty} f_p(\theta)\, d\theta = \gamma/2. \tag{3.14}$$

Analysis of the transformation of the subjective distribution shows that the a-posteriori distribution is obtained via Bayes' equation, taking into account only the experimental data. The third condition of rationality is thus satisfied.

It is easy to see that the updating of the information concerning the θ value may be done in one or several successive steps. Each step involves the application of the Bayes equation, taking as the a-priori distribution the a-posteriori distribution from the preceding step. In practice, correction of the a-priori opinion is a continuous process, as the experimental data are accumulated over time. The data may or may not agree with the initial opinion, based on the information available before the experiment. Considering a succession of experiments and taking as the a-priori distribution for experiment (i) the a-posteriori distribution from experiment $(i - 1)$, we get successive corrections of the initial opinion, the decision concerning the θ value coming closer and closer to reality. Of course, this will happen only if the initial range of the possible values of θ is not incorrectly truncated.

The decrease of the uncertainty associated with the θ value is described by measures of the variability. As the information is accumulated, the confidence interval and the variance decrease, and Shannon's and Fisher's measures of information, as well as the informational energy, increase. Asymptotically, an a-posteriori distribution concentrated around the actual value of θ is obtained, irrespective of the form of the a-priori distribution. Thus even if the initial information is erroneous, the final estimation will be correct because the experimental data overwhelm the initial opinion, on the condition already mentioned that the a-priori distribution should cover the entire range of θ. This reasoning should not lead to the uncritical adoption of any a-priori distribution. The bayesian estimation is surely useful in those situations when the amount of data is small. Therefore, a rational method to build up an a-priori distribution is essential. As this problem will be examined later (Section 3.2) we shall consider in the following some examples of a-priori distributions used in reliability theory.

The beta distribution is used in non-parametric bayesian estimation, when the character of the time-to-failure distribution is ignored. The reliability function plays the role of θ in the beta a-priori distribution:

$$f_a(R) = \frac{\Gamma(n_0)}{\Gamma(r_0)\,\Gamma(n_0 - r_0)} R^{n_0-r_0-1}(1 - R)^{r_0-1} \tag{3.15}$$

$n_0, r_0 > 0.$

The a-priori expected value of R is:

$$E_{a_R} = \int_0^1 R f_a(R)\,\mathrm{d}R = \frac{n_0 - r_0}{n_0}. \tag{3.16}$$

From equation (3.16), we can interpret the parameters n_0 and r_0 of the beta distribution as the numbers of pseudo-trials and pseudo-failures, respectively.

The a-priori variance of R is:

$$V_{a_R} = \int_0^1 R^2 f_a(R)\,\mathrm{d}R - E^2_{aR} = \frac{r_0(n_0 - r_0)}{n_0(n_0 + 1)}. \tag{3.17}$$

The relative uncertainty concerning the actual R value is measured by the degree of skewness of the a-priori distribution:

$$\frac{\sigma_{a_R}}{E_{a_R}} = \frac{\sqrt{r_0}}{\sqrt{(n_0 + 1)(n_0 - r_0)}}. \tag{3.18}$$

The capacity of the beta distribution to express the information available a-priori is illustrated in Fig. 3.2 by a family of beta distribution with mean values 0.5 and different levels of relative uncertainty. The beta distribution is symmetrical with respect to the mean value, and its variance decreases with the increase of r_0 and n_0. It can also be seen that, for r_0 and n_0 values less than unity, the beta distribution is U-shaped, a form which is seldom adequate to express the a-priori information. Therefore such values are to be avoided.

We shall show the transformation of the beta distribution under the impact of experimental data. The likelihood function is of the binomial type

$$f(x/R) = C_n^r (1 - R)^r R^{n-r} \tag{3.19}$$

where r is the number of failures and n the sample size.

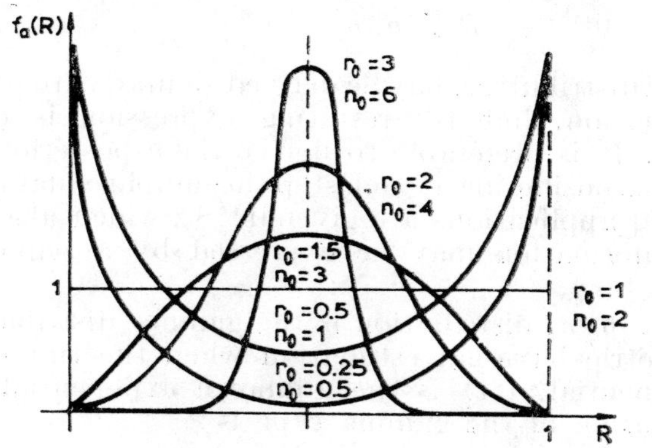

Fig. 3.2. — A family of beta distributions.

Introducing expressions (3.15) and (3.19) into the Bayes equation we get

$$f_p(R) = \frac{f_a(R)\, f(x/R)}{\int_0^1 f(x/R)\, f_a(R)\, \mathrm{d}R} =$$

$$= \frac{\Gamma(n_0 + n)\, R^{n_0+n-r-r_0-1}(1 - R)^{r+r_0-1}}{\Gamma(r_0 + r)\, \Gamma(n_0 + n - r_0 - r)}. \tag{3.20}$$

The a-posteriori distribution is also of the beta type, with parameters $n_0 + n$ and $r_0 + r$. The bayesian estimate of R is the posterior mean

$$\hat{R} = \int_0^\infty R f_p(R)\, \mathrm{d}R = \frac{n_0 + n - r_0 - r}{n_0 + n}. \tag{3.21}$$

The beta distribution is conservative with respect to the Bayes transformation if the distribution of the data is binomial. In [7] the necessary and sufficient conditions for the beta distribution to express the a-priori information are derived. These conditions are sufficiently loose that the decision to adopt a beta distribution remains a matter of choice.

For $n_0 = 2$ and $r_0 = 1$ the beta distribution becomes uniform, which means that no a-priori information is available on the R value. The a-posteriori distribution and the bayesian estimate for this case may be obtained from (3.20) and (3.21) for $n_0 = 2$, $r_0 = 1$. When only the range $[a, b]$ of R is known a-priori, for instance the possible values of R are limited to the interval (0.10 — 0.85), it is reasonable to adopt a uniform a-priori distribution:

$$f_a(R) = \begin{cases} \dfrac{1}{b - a} ; & R \in [a, b] \\ 0 ; & R \notin [a, b]. \end{cases} \tag{3.22}$$

The a-posteriori distribution may be derived similarly from (3.22), (3.19) and the Bayes equation, but the resulting expression is complicated and difficult to use. It is preferable to derive the a-posteriori distribution in successive steps, considering at each step the sample equal to unity. Numerical results and applications are given in [8], which also shows the way the a-priori information may be expressed by a sequence of uniform distributions.

Another a-priori distribution is the gamma distribution, frequently used in parametric bayesian estimation when the time-to-failure distribution is exponential. If λ is the unknown exponential parameter, the a-priori distribution of the gamma type is:

$$f_a(\lambda) = \frac{\beta(\beta\lambda)^{\alpha-1} e^{-\beta\lambda}}{\Gamma(\alpha)} \propto \lambda^{\alpha-1} e^{-\beta\lambda} ; \tag{3.23}$$

$$\alpha, \beta > 0.$$

The a-priori mean and variance are:

$$E_{a_\lambda} = \frac{\alpha}{\beta} ; \quad V_{a_\lambda} = \frac{\alpha}{\beta^2}. \tag{3.24}$$

The gamma distribution may express a variety of a-priori information because of its flexibility.

In order to describe the bayesian transformation of the gamma distribution we recall that, irrespective of the planning of the reliability test, the likelihood function of the data is of the form:

$$f(x/\lambda) = \text{constant } \lambda^r e^{-\lambda T_\Sigma} \propto \lambda^r e^{-\lambda T_\Sigma}. \tag{3.25}$$

From (3.23), (3.25) and the Bayes equation, we get the a-posteriori distribution:

$$f_p(\lambda) \propto \lambda^{\alpha+r-1} e^{-\lambda(\beta+T_\Sigma)}. \tag{3.26}$$

The expression (3.26) shows that the a-posteriori distribution is also of the gamma type, with parameters $\alpha + r$ and $\beta + T_\Sigma$. The gamma distribution is conservative in relation to the bayesian transformation if the time-to-failure is exponentially distributed.

The initial uncertainty about the actual λ value may be expressed by the variance (3.24) or by Fisher's information, which has the expression :

$$I_{Fa} = \int_0^\infty \left[\frac{\partial \ln f_a(\lambda)}{\partial \lambda} \right]^2 f_a(\lambda)\, d\lambda = \frac{\beta^2}{\alpha - 2}; \ \alpha > 2. \tag{3.27}$$

The concept of informational energy may also be used :

$$E_a = \int_0^\infty f_a^2(\lambda)\, d\lambda = \frac{\beta}{2^{2\alpha-1}} \frac{\alpha(\alpha+1)\ldots(2\alpha-2)}{\Gamma(\alpha)}. \tag{3.28}$$

From (3.24), (3.27) and (3.28) we notice that the uncertainty is low for large values of β and small values of α. The values of α and β must be strictly chosen according to the actual amount of information available a-priori, and care must be taken in order not to decrease artificially the uncertainty concerning the value of λ.

From the a-posteriori distribution (3.26), we derive the bayesian estimate of λ :

$$\hat{\lambda} = \mathrm{E}_{p\lambda} = \int_0^\infty \lambda f_p(\lambda)\, d\lambda = \frac{\alpha + r}{\beta + T_\Sigma}. \tag{3.29}$$

The bayesian confidence interval may easily be derived if α has an integer value. Otherwise, tables of the incomplete gamma function are necessary. In the integer case, we consider the new variable $v = 2(\beta + T_\Sigma)\lambda$, which is χ^2 distributed with $2(\alpha + r)$ degrees of freedom. From the tables of the chi-square distribution, we get the two-sided confidence interval :

$$\lambda_L = \frac{\chi^2_{\gamma/2}(2\alpha + 2r)}{2(\beta + T_\Sigma)}; \quad \lambda_U = \frac{\chi^2_{1-\gamma/2}(2\alpha + 2r)}{2(\beta + T_\Sigma)}. \tag{3.30}$$

The one-sided confidence interval is :

$$\lambda_U = \frac{\chi^2_{1-\gamma/2}(2\alpha + 2r)}{2(\beta + T_\Sigma)}. \tag{3.31}$$

The length of the confidence interval is a measure of precision. Comparing the classical one-sided confidence interval with the bayesian one, we notice that the precision is higher in the bayesian case and, for equal precisions, the experimental effort is reduced (prior information being available), by the quantity

$$\Delta T_\Sigma = \beta + T_\Sigma \left[1 - \frac{\chi^2_{1-\gamma}(2\alpha + 2r)}{\chi^2_{1-\gamma}(2r)} \right] \approx \beta. \tag{3.32}$$

6 – c. 531

Once the Bayes estimate of λ is obtained, any other reliability measure may be estimated provided that the relationship between this measure and the parameter λ is known. For instance, the reliability function $R(t) = e^{-\lambda t}$ may be estimated by taking the mean of $e^{-\lambda t}$ over the posterior distribution of λ.

A detailed analysis of the relationship between the precision and the cost of reliability tests, when prior information is available, is contained in [9] and [10]. A criterion of choice between type I and type II censored plans is developed, on the basis of the test costs and the loss function. For a quadratic loss function and a gamma prior, the type I censored plan is preferable. This result is contrary to the classical case, where the type II censored plan is cheaper.

A delicate problem in expressing the a-priori information is that the same level of uncertainty may be associated with all reliability measures like the failure rate, the mean life or the reliability function. If we adopt a gamma prior for λ, the values α and β express the relative uncertainty concerning the λ value. The same uncertainty should be associated with the value of the mean life, m. The a-priori distribution for m may not be chosen to be of the gamma type, because of the deterministic relationship between m and λ. The a-priori distribution of m, when a gamma prior is chosen for λ may be derived as follows:

$$f_{a_m}(m) = \frac{f_{a_\lambda}(\lambda)}{\left|\dfrac{dm}{d\lambda}\right|} = \frac{\beta(\beta\lambda)^{\alpha-1}\, e^{-\beta\lambda}}{\Gamma(\alpha)\dfrac{1}{\lambda^2}} = \frac{\beta^\alpha\, m^{-\alpha-1}\, e^{-\frac{\beta}{m}}}{\Gamma(\alpha)}. \tag{3.33}$$

The expression (3.33) is the inverted gamma distribution, with mean and variance:

$$E_{a_m} = \frac{\beta}{\alpha-1}; \quad V_{a_m} = \frac{\beta^2}{(\alpha-1)^2(\alpha-2)}; \quad \alpha > 2. \tag{3.34}$$

From (3.34) we notice that a low uncertainty on λ(large β, small α) implies a high uncertainty on m, which contradicts the fact that we know m to the same extent as we know λ. The contradiction may be eliminated when relative uncertainty is considered, as defined by (3.18). The relative uncertainties associated with λ and m are:

$$\begin{aligned} \frac{\sigma_{a_\lambda}}{E_{a_\lambda}} &= \frac{\sqrt{\dfrac{\alpha}{\beta^2}}}{\dfrac{\alpha}{\beta}} = \frac{1}{\sqrt{\alpha}}; \\ \frac{\sigma_{a_m}}{E_{a_m}} &= \frac{\sqrt{\dfrac{\beta^2}{(\alpha-1)^2(\alpha-2)}}}{\dfrac{\beta}{\alpha-1}} = \frac{1}{\sqrt{\alpha-2}}; \quad \alpha > 2. \end{aligned} \tag{3.35}$$

Concluding, the same a-priori information must be differently expressed for each reliability measure by different priors which are not independent. Let us take for instance the case when no information is available on the reliability function. A uniform prior for R is then adopted:

$$f_{a_R}(R) = \begin{cases} 1 & R \in [0,1]; \\ 0 & R \notin [0,1]. \end{cases} \tag{3.36}$$

The same ignorance must be attributed to λ, if the time-to-failure is exponentially distributed. Yet the a-priori distribution associated with λ is not uniform. Taking into account that $\lambda = (1/t)\ln(1/R)$, we get:

$$f_{a_\lambda}(\lambda) = \frac{f_{a_R}(R)}{\left|\dfrac{d\lambda}{dR}\right|} = tR = te^{-\lambda t} \tag{3.37}$$

which is a gamma distribution with $\alpha = 1$ and $\beta = t$. The relative uncertainties on λ and R are equal, but the a-priori distributions are different.

With these remarks, we end the presentation of the methodology of bayesian estimation. This methodology allows a rational combination of the a-priori and experimental information and leads to easy computations if conservative priors are used. Correct use of the bayesian estimation implies a considerable reduction of the amount of data necessary to attain a reasonable precision. The value of this methodology is conditioned by the way the a-priori information is converted into a subjective distribution, without modifying the actual degree of uncertainty.

The a-priori distribution is subjective only in the sense of the subjective interpretation of probability, and it must reflect the information objectively available. Otherwise, the bayesian decisions are erroneous and possibly in contradiction with further observations.

3.2 Expressing a-priori information

The information concerning a parameter of the reliability model, that is available before performing any experiment, may be expressed in terms of a subjective pdf associated with the unknown parameter which should cover the entire range of the parameter values. It is desirable that the a-priori distribution should be unique, that is, equal amounts of information should lead to the same a-priori distribution irrespective of the analyst.

The information theory approach provides a method for deriving the a-priori distribution, using the entropy as a measure of the uncertainty on the parameter to be estimated. The entropy is defined by

$$H = -\int_0^\infty f_a(\lambda) \log f_a(\lambda)\, d\lambda. \tag{3.38}$$

The a-priori distribution must be so chosen that the entropy (3.38) is maximized while satisfying the constraints imposed by the prior knowledge. This procedure is of the "worst case" type, and it is intended to prevent the analyst from quantifying his desires rather than the objective information in constructing the a-priori distribution.

If no prior information is available, the a-priori distribution which maximizes the entropy is uniform. The principle of maximum entropy is then equivalent to the principle of insufficient reason of Laplace, which states that two events are equally probable if there is no evidence to consider them otherwise.

In order to apply the principle of maximum entropy, the a-priori information must be expressed in terms of the expectations of K functions g_i of the parameter to be estimated. These expectations are the constraints imposed on the a-priori distribution :

$$\int_0^\infty g_i(\lambda) f_a(\lambda)\, \mathrm{d}\lambda = \bar{g}_i; \quad i = 1,2, \ldots, K. \tag{3.39}$$

The a-priori distribution $f_a(\lambda)$ is uniquely determined by maximizing the entropy (3.38), subject to the constraints (3.39) and to the conditions that $f_a(\lambda)$ should represent a probability density function :

$$f_a(\lambda) \geqslant 0; \quad \int_0^\infty f_a(\lambda)\, \mathrm{d}\lambda = 1. \tag{3.40}$$

We have already solved a similar problem in Chapter 2, by the method of Lagrange multipliers.

For $K = 1$, the a-priori distribution is :

$$f_a(\lambda) = \frac{\mathrm{e}^{-\beta g(\lambda)}}{\int_0^\infty \mathrm{e}^{-\beta g(\lambda)}\, \mathrm{d}\lambda} \tag{3.41}$$

where β is determined from :

$$\frac{\mathrm{d}}{\mathrm{d}\beta}\left[\ln \int_0^\infty \mathrm{e}^{-\beta g(\lambda)}\, \mathrm{d}\lambda\right] = -\bar{g}. \tag{3.42}$$

When $K > 1$ the expectations $\bar{g}_1, \bar{g}_2, \ldots, \bar{g}_K$ are known, and the a-priori distribution is

$$f_a(\lambda) = \frac{\mathrm{e}^{-\sum_{i=1}^{K} \beta_i g_i(\lambda)}}{\int_0^\infty \mathrm{e}^{-\sum_{i=1}^{K} \beta_i g_i(\lambda)}\, \mathrm{d}\lambda} \tag{3.43}$$

where the constants β_i are determined from the equations

$$\frac{\partial}{\partial \beta_i}\left[\ln \int_0^\infty e^{-\sum_{i=1}^{K} \beta_i g_i(\lambda)} \mathrm{d}\lambda\right] = -\bar{g}_i; \quad i = 1,2,\ldots,K. \tag{3.44}$$

Thus, the widest a-priori distribution compatible with the available information is uniquely specified. The major difficulty in applying this method lies in expressing the a-priori information in terms of the expectations of the functions $g_i(\lambda)$.

Let us consider the case when $K = 1$ and $g(\lambda) = \lambda$, that is the expected a-priori value of the parameter is known

$$\lambda_a = \int_0^\infty \lambda f_a(\lambda)\, \mathrm{d}\lambda. \tag{3.45}$$

From relation (3.41) we get an a-priori distribution of the exponential type :

$$f_a(\lambda) = \frac{e^{-\beta\lambda}}{\int_0^\infty e^{-\beta\lambda}\, \mathrm{d}\lambda} = \beta e^{-\beta\lambda}. \tag{3.46}$$

The parameter β of the exponential distribution (3.46) may be computed according to (3.42) :

$$\frac{\mathrm{d}}{\mathrm{d}\beta}\left[\ln \int_0^\infty e^{-\beta\lambda}\, \mathrm{d}\lambda\right] = -\lambda_a \tag{3.47}$$

which implies $\beta = \dfrac{1}{\lambda_a}$.

The parameter of the a-priori distribution is thus equal to the expected mean time-to-failure before the experiment. The fact that the expected value of the parameter is a-priori known may be interpreted in different ways. Strictly speaking, knowledge of λ_a implies the possibility of measuring it with sufficient accuracy, so that the uncertainty about it is negligible [12]. In this case there is no point in further estimation, as the parameter is perfectly known from the beginning.

If the prior information consists in recognizing the existence of the prior expectation without knowing precisely its value, then the experimental data are bound to confirm, or not, the expected value of the parameter, by bayesian inference. The difficulty arises from the fact that there are numerous functions of the parameter which have prior mean values, and one could not use them all as constraints in deriving the a-priori distribution. It is then necessary to give a selection rule which should be as objective as possible. Moreover, even the formulation of the prior information in terms of the expectations of some functions of the parameter is often cumbersome. The a-priori distribution which results by applying the maximum entropy principle is not always conservative with respect to the Bayes transform, and it is therefore difficult to use. This explains the critical point of view expressed in [12, 13].

The limited possibility of applying a general principle like that of maximum entropy brings forth a certain degree of subjectivity in deriving the a-priori distribution. Different analysts using the same amount of information would obtain different a-priori distributions. However, it is expected that these distributions would have the same measure of uncertainty. The simplest solution in this case is to adopt a conservative distribution (the natural conjugate) like the beta or gamma distributions. The parameters of the a-priori distribution must be specified according to the actual uncertainty on the value of the parameter.

Let us assume that a gamma distribution is adopted as the a-priori distribution associated with the parameter of the exponential distribution of the time-to-failure. The a-priori information comprises the values of the constant failure rate of similar systems and, sometimes, the results of reliability tests performed on similar systems, that is the number of failures r and the total time on test, T_Σ. When only the value of the failure rate is known, it is adopted as the a-priori mean :

$$E_{a_\lambda} = \frac{\alpha}{\beta} = \lambda_a . \tag{3.48}$$

The relative uncertainty on λ is measured by the skewness of the gamma distribution, which is equal to $1/\sqrt{\alpha}$. A reasonable value of the skewness may be adopted, taking into account the degree of similarity between the system which has the failure rate λ_a and the system to be analysed. The value of the skewness gives the parameter α, and relation (3.48) may be used to compute the parameter β.

When the experimental data on the similar system are also known, the parameters of the gamma distribution may be computed according to the relations :

$$\alpha = kr\,; \quad \beta = kT_\Sigma. \tag{3.49}$$

The coefficient k in relations (3.49) ranges over the interval (0,1) and depends on the relevance of the a-priori information in the case of the system considered. In [5], the values of k are selected according to Table 3.1.

The procedure is reasonable because the expected a-priori failure rate is equal to the maximum likelihood estimator, and the importance of the a-priori information is easily controlled by the value of the coefficient k. As the a-priori distribution is the conjugate of the time-to-failure distribution, the a-posteriori distribution is also of the gamma type, with the parameters shifted by the number of failures and the total time on test observed for the system at hand. The value of k specifies the weight of the data, obtained from reliability tests performed on similar systems, with respect to the data observed for the system considered.

Table 3.1

Specifying the a-priori distribution

k	Source of information
0.1	similar system, similar application
0.3	identical system, similar application
0.5	similar system, identical application
1	identical system, identical application

The same idea of comparing the amount of a-priori information to the amount of experimental information may be carried out using, as a common measure, the information in Fisher's sense [16]. Fisher's information of the gamma a-priori distribution I_{F_a} was previously computed and has the form (3.27).

The probability density function for the experimental data is $f(x/\lambda)$. If this function is flat with respect to λ the estimation cannot be accurate, and hence the information on λ is low. On the contrary, if $f(x/\lambda)$ is peaked with respect to λ, with steep slopes, the parameter can be accurately localized and the information is ample. It is then convenient to define Fisher's experimental information by the average of the squared slope of $\ln f(x/\lambda)$ with respect to λ:

$$I_{F_x} = \int_0^\infty \left[\frac{\partial \ln f(x/\lambda)}{\partial x} \right]^2 f(x/\lambda) \, \mathrm{d}x. \tag{3.50}$$

It is important to notice that, in relation (3.50), the mean is taken over all possible experimental data while, when computing the a-priori information, the mean is taken over the possible values of the parameter which is interpreted as a subjective random variable.

As we have shown in Chapter 2, the experimental information is synthesized by the total time on test T_Σ, which has the pdf:

$$f(T_\Sigma/\lambda) = \frac{\lambda(\lambda T_\Sigma)^{r-1}\, \mathrm{e}^{-\lambda T_\Sigma}}{\Gamma(r)}. \tag{3.51}$$

Introducing expression (3.51), we get Fisher's experimental information:

$$I_{F_x} = \frac{r}{\lambda^2}. \tag{3.52}$$

Fisher's information depends on the unknown value of λ. Using the maximum likelihood estimation of λ, Fisher's information may be estimated as:

$$\hat{I}_{F_x} = \frac{r}{\hat{\lambda}^2}. \tag{3.53}$$

The ratio K between the a-priori and the experimental information may be initially specified, so that:

$$I_{F_a} = K I_{F_x}. \tag{3.54}$$

The ratio K is at most equal to unity, when the a-priori information is as important as the experimental. The choice of actual value of K must be made according to the relevance of the a-priori information. Using expressions (3.27) and (3.53), equation (3.54) becomes:

$$\frac{\beta^2}{\alpha - 2} = K \frac{r}{\hat{\lambda}^2}. \tag{3.55}$$

If the expected a-priori value of λ is λ_a, another equation with solutions α and β may be written:

$$\frac{\alpha}{\beta} = \lambda_a. \tag{3.56}$$

The parameters of the a-priori distribution are completely determined by solving equations (3.55) and (3.56).

Although Fisher's a-priori information is estimated from the data, that is the data are used in part to determine the a-priori distribution, this does not mean changing the prior after observing the data because the relative amount of information distributed between the prior and the data is determined by the K value before observing the data.

Assume, as a numerical example, that the expected a-priori failure rate is 10^{-4}h^{-1}. The a-priori information is 25% of the experimental information. The experiment is censored at $r = 2$ and the total time on test is $T_\Sigma = 10^5$ h. A Bayes estimate of the failure rate is required. First, the unbiased maximum likelihood estimator of λ is computed from the data:

$$\hat{\lambda} = \frac{r-1}{T_\Sigma} = 10^{-5}\text{h}^{-1}.$$

Fisher's information provided by the experiment is estimated:

$$\hat{I}_{F_x} = \frac{r}{\hat{\lambda}^2} = 2 \cdot 10^{10}\,\text{h}^2.$$

A gamma a-priori distribution with parameters α and β is adopted. The a-priori Fisher's information is $\beta^2/(\alpha - 2)$, and is equal to 1/4 of the experimental information :

$$\frac{\beta^2}{\alpha - 2} = 5 \cdot 10^9 \text{ h}^2.$$

The prior expected value of λ is α/β, so that :

$$\frac{\alpha}{\beta} = 10^{-4} \text{ h}^{-1}.$$

Solving the equations in α and β, we get :

$$\alpha = 50; \quad \beta = 5 \cdot 10^5 \text{ h}.$$

The a-posteriori distribution is also of the gamma type, with parameters $\alpha + r$ and $\beta + T_\Sigma$. The Bayes estimate of λ is the posterior mean :

$$\hat{\lambda}_B = \frac{\alpha + r}{\beta + T_\Sigma} = \frac{52}{6 \cdot 10^5} = 8.66 \cdot 10^{-5} \text{ h}^{-1}.$$

The method described above uses the data in order to estimate the parameters of the a-priori distribution. This is apparently wrong, as the a-priori information must be independent of the experiment, both being blended into the a-posteriori distribution. Correction of the initial opinion is the effect of the bayesian inference, so that it is not reasonable to modify the a-priori distribution after observing the data. Actually, the independence of the a-priori and experimental information is assured by specifying their relative weights before performing the experiment. The data are then used only to estimate Fisher's experimental information. Moreover, the a-priori expected value λ_a need not be equal to the maximum likelihood estimate of $\hat{\lambda}$, so that there is not any retrospective modification of the a-priori opinion.

Before the final adoption of the a-priori distribution, it is necessary to test its behaviour when confronted with hypothetical data which confirm or contradict the initial opinion. Such an analysis is called preposterior, because it involves computation of the a-posteriori distribution by introducing the candidate for the a-priori distribution and the simulated data into the Bayes equation. It is desirable that the a-posteriori and a-priori distributions should clearly differ when the data contradict the initial opinion, but be only slightly different when the data support the initial opinion. The a-priori distribution must be sensitive to the impact of the data which contradict the initial opinion and relatively little affected (conservative) when the data agree with the initial opinion.

Should the a-priori distribution be over-sensitive with respect to the data which confirm the a-priori information, the Bayes estimate would be excessively optimistic, the uncertainty would decrease disproportionately to the restricted volume of the experiment, and the prediction would ultimately fail.

Conversely, the a-priori distribution must be sensitive when few data contradict the initial opinion, thus accounting for the adaptive feature of the bayesian inference.

During the pre-posterior analysis, data are simulated which confirm or contradict the initial opinion, and the a-posteriori distribution is computed for each data set. If the conditions mentioned above are fulfilled, the a-priori distribution may finally be adopted and combined with the actual data.

Such a pre-posterior analysis is carried out in [8] using an a-priori beta distribution with parameters r_0 and n_0. Let $r_0 = 1$ and $n_0 = 100$. This means that the prior expected value of the reliability function is 0.99 and that a low uncertainty is associated with this statement. A sample of $n = 10$ is simulated so that no failures are observed during the test time, that is $r = 0$. Such data confirm the initial opinion of a high reliability. The posterior distribution is also of the beta type, with parameters $n_0 + n = 110$ and $r_0 + r = 1$, which means a Bayes estimate of the reliability function equal to $109/110 = 0.991$. The a-priori distribution is then conservative when the data agree with the a-priori information.

When 10 successive failures are simulated, which means that the initial opinion is severely amended, a posterior distribution of the beta type with parameters $n_0 + n = 110$ and $r_0 + r = 11$ is obtained, which leads to a Bayes estimate of the reliability function equal to 0.90. This result is unsatisfactory because the analyst still believes in a high value of the reliability function, in spite of the ten failures recorded. The beta distribution with parameters $r_0 = 1$ and $n_0 = 100$ behaves satisfactorily when the data agree with the prior information, but fails to adapt itself to the data which contradict it. The reason for this behaviour is the low prior uncertainty associated with the parameter value.

Now let the parameters of the beta distribution be $r_0 = 0.01$ and $n_0 = 1$. This implies the same a-priori expected reliability function of 0.99 but a higher uncertainty associated with it. When this initial opinion is combined with 10 successive failures, the Bayes estimate of the reliability function is $0.99/11 = 0.09$, which is reasonable because the initial opinion is severely amended. On the other hand, when 10 successes occur, the Bayes estimate of the reliability function is $(10 + 0.99)/(1 + + 10) = 0.9991$, which represents an order of magnitude increase in the posterior distribution mean compared to the prior estimate and again does not fully meet the desired behavioural desiderata. It is true that the Bayes estimate is reasonably situated between the maximum likelihood estimate (which is unity) and the prior expected value, but other difficulties arise in the computation of the confidence interval. Thus, for small values of the parameters, the beta distribution is U-shaped with fractions r_0/n_0 of the probability assigned near the value $R = 1$ and $(n_0 - r_0)/ /n_0$ near the value $R = 0$, with almost nothing assigned between. This might result in a prior with mean of $R = 0.95$ having a bayesian 90% lower confidence bound on R higher than the mean value as a result of this property of extreme aggregation. The confidence interval is difficult to interpret reasonably [8].

In summary, the beta a-priori distribution with small values of the parameters behaves well in the sense that it is annihilated by moderate length test sequences containing failures. When no failures are encountered, the Bayes estimate of the reliability is excessively high.

A similar pre-posterior analysis may be made for the gamma a-priori distribution. Let us assume that the experimental data confirm the initial opinion, that is the maximum likelihood estimate is approximately equal to the a-priori expected value of the parameter λ :

$$\frac{r}{T_\Sigma} \approx \frac{\alpha}{\beta} = \lambda_a . \tag{3.57}$$

The Bayes estimate is the mean posterior, that is :

$$E_{p_\lambda} = \frac{\alpha + r}{\beta + T_\Sigma} \approx \frac{\alpha}{\beta} = \lambda_a . \tag{3.58}$$

The ratio of the relative uncertainties associated with the posterior and the prior distributions is :

$$\frac{\sigma_{p_\lambda}}{\sigma_{a_\lambda}} = \sqrt{\frac{\alpha}{\alpha + r}} . \tag{3.59}$$

From (3.58) and (3.59), it may be seen that the gamma distribution is conservative when the data confirm the initial opinion. The Bayes estimate is close to the prior expected value and the uncertainty decreases moderately.

Let us consider a data set which contradicts the initial opinion, such that the maximum likelihood estimator is two orders of magnitude higher than the a-priori mean :

$$\frac{r}{T_\Sigma} = 10^2 \frac{\alpha}{\beta} . \tag{3.60}$$

The Bayes estimate is in this case :

$$E_{p_\lambda} = \frac{\alpha + r}{\beta + T_\Sigma} = \frac{\beta + 10^2\, T_\Sigma}{\beta + T_\Sigma} \cdot \frac{\alpha}{\beta} =$$

$$= \frac{\alpha + r}{10^2 \alpha + r} \cdot 10^2 \frac{\alpha}{\beta} . \tag{3.61}$$

From (3.61) it can be seen that the gamma distribution is sufficiently sensitive when the data contradict the initial opinion, unless either of the following inequalities are true :

$$\alpha \gg r \quad \text{or} \quad \beta \gg T_\Sigma . \tag{3.62}$$

As a rule, the unknown failure rate ranges in the interval (10^{-4} — — 10^{-7} h^{-1}). If α is greater than unity, β must have a high value, which could imply a low sensitivity according to relation (3.62). On the other hand, adopting a small value for α would imply a high initial uncertainty, so that the a-priori information would easily be wiped out by the data, and there would be little point in using it.

The difficulties involved in the use of a conjugate distribution, as pointed out by the pre-posterior analyses performed, may be avoided by using a "free hand" a-priori distribution which would be a fair portrayal of the decision maker's judgement, however reached. Such a technique has obvious difficulties in terms of mathematical tractability; however, it does suggest an alternative technique of using discrete a-priori distributions which share the mathematical simplicity of the continuous conjugate and offer almost infinite choice of distributional shape [8].

The establishment of an appropriate discrete prior cannot be satisfactorily accomplished by individuals who are uninformed or unqualified in regard to the equipment or the reliability measure (usually the reliability function or the failure rate) under consideration. This a-priori distribution should be chosen by a panel of design engineers who are well versed in the equipment, familiar with its developmental history and performance, and familiar with the reliability characteristics of similar or related equipment in similar or related applications.

It is necessary that the a-priori distribution should cover the entire range of the reliability measure. The complete range is divided into several cells of unequal length. The cells are more numerous at higher reliability values, because there a sharp discrimination (between say $R = 0.999$ and $R = 0.99$) may have real meaning whereas the distinction between $R = 0.51$ and $R = 0.50$ is not so important. Prior probability assignments are then made for each cell, in accordance with the group's strength of belief in each hypothesis. The pre-posterior analysis shows a satisfactory behaviour of the discrete a-priori distribution with respect to different possible data sets. The major difficulty in the development of this a-priori distribution is the combining of individual opinions so that the probabilities assigned to each cell should represent the opinion of the group.

The criteria to combine individual opinions into the opinion of the group are presented in [19]. Let θ denote the reliability measure. Each individual expresses his opinion on θ by an a-priori distribution $f_{a_i}(\theta)$ or by an expected value of the parameter θ_{a_i}. This expected value is the mean of the a-priori distribution, as the members of the group are not yet faced with the experimental data. After the data are recorded, there are two possible procedures. In the first, the individual opinions may be updated in the posterior distributions $f_{p_i}(\theta)$, and the posterior means θ_{p_i} are considered. A second procedure is to use in the bayesian inference the a-priori distribution associated with the group, which has previously been derived from the individual pdfs $f_{a_i}(\theta)$. The first procedure has the disadvantage that the Bayes individual estimators θ_{p_i} are dependent, as each member of the group is using the same experimental data.

Let $\hat{\theta}_g$ denote either the prior or the posterior estimate of the group. It is a function of the individual estimates $\hat{\theta}_1, \hat{\theta}_2, \ldots, \hat{\theta}_N$.

$$\hat{\theta}_g = g(\hat{\theta}_1, \hat{\theta}_2, \ldots, \hat{\theta}_N). \tag{3.63}$$

The function g must be consistent, that is, it must not decrease as one member of the group increases his estimate. Secondly, the function g should be such that the group estimate could not be totally determined by some subgroup. This condition also excludes a dictatorship in which one person totally determines a group's prediction.

The substantive condition states that the nature of $\hat{\theta}_g$ should not be different from the nature of the individual $\hat{\theta}_i$ values. Thus, if the $\hat{\theta}_i$ are probabilities (e.g. values of the reliability function), then $\hat{\theta}_g$ must also be a probability.

A last condition refers to the performances of the group estimator with respect to those of the individual estimators. The performance is related to the estimation accuracy as described, for instance, by the variance of the estimator. The accuracy of the group estimator $\hat{\theta}_g$ should be superior to the accuracy of any individual estimator.

A function which satisfies these conditions is the weighted average [19] :

$$\hat{\theta}_g = \sum_{i=1}^{N} w_i \hat{\theta}_i. \tag{3.64}$$

The weights w_i are determined by minimizing the variance of $\hat{\theta}_g$. The values of the weights are :

$$w_i = \frac{1}{\sigma_i^2 \sum_{j=1}^{N} \frac{1}{\sigma_j^2}} \tag{3.65}$$

where σ_i^2 is the variance of the individual estimate $\hat{\theta}_i$. When the weights are chosen according to (3.65), the minimum variance of $\hat{\theta}_g$ is :

$$\sigma_g^2 = \frac{1}{\sum_{i=1}^{N} \frac{1}{\sigma_i^2}} \,. \tag{3.66}$$

It is obvious that the weight of an individual estimate is higher when its variance is lower, that is, the uncertainty associated with the estimate is lower. It is natural that a less uncertain opinion should be more important, but the members of the group should not artificially underrate their uncertainty in order to gain a more prominent position in the group.

As stated before, relations (3.64) and (3.65) may combine either the prior or the posterior individual estimates. However, it seems better to combine the prior estimates, and then update the group estimate by the Bayes inference, because the individually updated predictions are not independent.

The method described above may be employed in the analysis of systems. In a global approach, the individual opinions on the system as a whole are combined, while in a structural approach, the individual opinions on the components are combined and thereafter a structural analysis performed. These two procedures are not equivalent when the weighted average is employed in deriving the group's estimate.

Let us consider the simplified case of a system with two components, so that the system fails when either component fails. The group is composed of two analysts who have the same amount of information on the reliability of the system. Let $\hat{R}_{ij}$ denote the estimate of analyst i of the reliability function of component j. The individual estimates of the reliability of the system are:

$$\hat{R}_1 = \hat{R}_{11} \cdot \hat{R}_{12}\,; \quad \hat{R}_2 = \hat{R}_{21} \cdot \hat{R}_{22}. \tag{3.67}$$

As the uncertainties of the analysts are equal, the group estimate of the system reliability is:

$$\hat{R}_g = \frac{\hat{R}_1 + \hat{R}_2}{2} = \frac{1}{2}\,(\hat{R}_{11} \cdot \hat{R}_{12} + \hat{R}_{21} \cdot \hat{R}_{22}). \tag{3.68}$$

If the estimates of the component reliabilities are combined, the group estimates of the component reliabilities are:

$$\hat{R}_{g_1} = \frac{\hat{R}_{11} + \hat{R}_{21}}{2}\,; \quad \hat{R}_{g_2} = \frac{\hat{R}_{21} + \hat{R}_{22}}{2}\,. \tag{3.69}$$

A simple structural analysis gives the reliability function of the system as

$$\hat{R}_g = \hat{R}_{g_1}\,\hat{R}_{g_2} = \frac{1}{4}\,(\hat{R}_{11} + \hat{R}_{21})\,(\hat{R}_{21} + \hat{R}_{22}). \tag{3.70}$$

The comparison of relations (3.68) and (3.70) shows that the two procedures are not equivalent.

It is preferable to combine the individual opinions on the components to form group estimates of the component reliabilities and then to compute the system reliability, because the accuracy of the final estimator is higher [19].

This Section has described the methods employed to convert into numerical form the information available on the reliability of the system before performing any life test. This information is expressed by a distribution of subjective probabilities associated with the unknown value of the reliability measure. The present state of the art does not allow a unique determination of the a-priori distribution on the basis of the amount of information available. In the absence of a general method of deriving the a-priori distribution, the amount of a-priori information was compared to the amount of experimental information in order to select an a-priori distribution among a family of convenient pdfs. The final adoption of the a-priori distribution was preceded by a pre-posterior analysis which studied the behaviour of the a-priori distribution under the impact of the experimental data. This analysis showed the advantage of the discrete a-priori distribution, based on the combining of the opinions of the members of a group.

In the tendency to increase the objectivity of the a-priori information, this was restricted to past experimental data. Consequently, a strictly objective interpretation of the bayesian inference was developed, as will be shown in the next Section.

3.3 Empirical Bayes methods

The impossibility of finding a unique general method, based on a certain amount of information, for determining the a-priori distribution, led to the development of several methods affected more or less by the subjectivity of the analyst.

The radical removal of this subjectivity may be achieved by restraining the available information to the objective results of past experiments. The previous data could be accumulated and used in a classical statistical inference if the parameters of the time-to-failure distribution did not actually vary from experiment to experiment. As a matter of fact, the empirical Bayes estimation is a method of pooling life test data and thus accounting for the random character of the failure rate. Generally, an empirical Bayes method provides an empirical means of pooling all the parameter estimates in such a way that the resulting estimate of the parameter in the current experiment has a smaller mean squared error than the estimate obtained without empirical Bayes-pooling.

The fact that the parameter θ of the time-to-failure distribution varies from experiment to experiment is modelled by an a-priori distribution $f_a(\theta)$ associated with θ. Unlike the a-priori distribution employed in bayesian inference, this a-priori distribution is interpreted in an objective way, that is, the value of the parameter θ corresponding to the data in hand is determined by sampling from a population with a distribution $f_a(\theta)$. The a-priori distribution is combined with the current experimental data by the Bayes equation, and the resulting a-posteriori distribution gives the point and interval estimates of θ. The procedure of updating the estimates when new experimental data are encountered is different from the case when the a-priori distribution is subjectively interpreted. If one is performing a sequence of experiments and if θ does vary from experiment to experiment according to $f_a(\theta)$, then it is not proper to use the posterior from one experiment as the prior for the next, since this posterior distribution is not the one generating the next value of θ. On the other hand, if $f_a(\theta)$ represents one's belief, before starting the sequence of experiments, about the single θ which will obtain throughout the sequence, then it is correct bayesian practice to use the posterior from one experiment as the prior to the next to describe how one's personal probability is modified.

In the case of empirical Bayes estimation, the experimental data are generated according to the combined action of two probabilistic laws : the time-to-failure distribution, and the a-priori distribution associated with its parameter. Hence, the marginal distribution of the data may be derived by the formula of total probability [21]. The parameters of the marginal distribution may be estimated by classical methods from the data obtained in the sequence of experiments and, consequently, the parameters of the prior are estimated through observations taken from the marginal distribution. Testing hypotheses on the prior distribution may be performed by confronting the marginal distribution with the data. The power of such a test is low, so that it is difficult to determine even the general shape of the prior on the basis of observed data from the marginal distribution, even if the data sets are moderately large.

However, if a flexible a-priori distribution is specified parametrically from a sufficiently large family, it is convenient to select it by estimating its parameters without performing any goodness-of-fit test.

The gamma distribution is used in [22] to describe the random variation of the parameter λ of the exponential distribution of the time-to-failure. Suppose that the available information consists of the data obtained in K type II censored samples at r failures, so that the last sample corresponds to the current experiment. Let t_{ij} be the time-to-failure of the element j in the sample i and n_i the sample size. The information obtained in an experiment is synthesized by the total time on test, which is

$$T_{\Sigma_i} = \sum_{j=1}^{r} t_{ij} + (n_i - r)\, t_{ir} \tag{3.71}$$

in the non-replacement case, and

$$T_{\Sigma_i} = n_i\, t_{i_r} \tag{3.72}$$

when the failed elements are replaced by new ones.

The pdf of the total time on test is in both cases of the gamma type with parameters r and λ :

$$f(T_\Sigma/\lambda) = \frac{\lambda(\lambda T_\Sigma)^{r-1}\, e^{-\lambda T_\Sigma}}{(r-1)!} \,. \tag{3.73}$$

The a-priori distribution associated with the random parameter is a gamma pdf with parameters α and β :

$$f_a(\lambda) = \frac{\beta(\beta\lambda)^{\alpha-1}\, e^{-\beta\lambda}}{\Gamma(\alpha)} \,. \tag{3.74}$$

The marginal distribution of the total time on test may be derived by introducing expressions (3.73) and (3.74) in the formula of total probability :

$$f(T_\Sigma) = \int_0^\infty f(T_\Sigma/\lambda) f_a(\lambda)\,\mathrm{d}\lambda = \frac{\Gamma(\alpha + r)}{\Gamma(\alpha)\,\Gamma(r)} \cdot \frac{\beta^\alpha\, T_\Sigma^{r-1}}{(\beta + T_\Sigma)^{\alpha+r}}. \tag{3.75}$$

The marginal distribution (3.75) resulted from the combination of two random factors : on the one hand the random character of the time-to-failure which accounts for the dissimilarity of the elements in a sample, and on the other hand the random character of the parameter λ, which accounts for the non-homogeneity of the samples.

The parameters α and β may be estimated by the method of moments. The first two moments of T_Σ are :

$$E_{T_\Sigma} = \int_0^\infty T_\Sigma f(T_\Sigma)\,\mathrm{d}T = \frac{\beta r}{\alpha - 1}; \quad \alpha > 1$$

$$E_{T_\Sigma^2} = \int_0^\infty T_\Sigma^2 f(T_\Sigma)\,\mathrm{d}T_\Sigma = \frac{\beta^2 r}{(\alpha - 1)(\alpha - 2)}. \tag{3.76}$$

The theoretical moments (3.76) are set equal to the respective empirical moments, so that the equations of the estimates of α and β are obtained :

$$m_1 = \frac{\sum_{i=1}^{K} T_{\Sigma_i}}{K} = \frac{\beta r}{\alpha - 1};$$

$$m_2 = \frac{\sum_{i=1}^{K} T_{\Sigma_i}^2}{K} = \frac{\beta^2 r(r+1)}{(\alpha - 1)(\alpha - 2)}. \tag{3.77}$$

Solving equations (3.77) we get :

$$\hat{\alpha} = \frac{m_1^2(r+1) - 2m_2}{m_1^2(r+1) - m_2}; \quad \hat{\beta} = \frac{(\hat{\alpha} - 1)\, m_1}{r}. \tag{3.78}$$

In practical situations it is difficult to have the samples censored at the same number of failures. Relations (3.77) may be modified to accommodate the case when the samples are censored at different failure numbers r_i. First, the relations (3.77) are written, in the equivalent form :

$$\frac{1}{K}\sum_{i=1}^{K} \frac{T_{\Sigma_i}}{r} = \frac{\beta}{\alpha - 1}; \quad \frac{1}{K}\sum_{i=1}^{K} \frac{T_{\Sigma_i}^2}{r(r+1)} = \frac{\beta^2}{(\alpha-1)(\alpha-2)}. \tag{3.79}$$

The right-hand sides of equations (3.79) are independent of r. Consequently, one may write :

$$m_1^* = \frac{1}{K} \sum_{i=1}^{K} \frac{T_{\Sigma_i}}{r_i} = \frac{\beta}{\alpha - 1};$$

$$m_2^* = \frac{1}{K} \sum_{i=1}^{K} \frac{T_{\Sigma_i}^2}{r_i(r_i + 1)} = \frac{\beta^2}{(\alpha - 1)(\alpha - 2)}. \tag{3.80}$$

Solving equations (3.80), we get :

$$\hat{\alpha} = \frac{2m_2^* - m_1^{*2}}{m_2^* - m_1^{*2}}; \quad \hat{\beta} = (\hat{\alpha} - 1)m_1^*. \tag{3.81}$$

Let us now consider a sequence of K type I censored samples having the same total time on test T_Σ. The number of failures in a sample is distributed according to the Poisson law :

$$f(r/\lambda) = \frac{(\lambda T_\Sigma)^r}{r!} e^{-\lambda T_\Sigma}. \tag{3.82}$$

The marginal distribution of the number of failures is obtained by introducing the distribution (3.82) and the a-priori distribution (3.74) into the formula of total probability :

$$f(r) = \int_0^\infty f(r/\lambda)\, f_a(\lambda) d\lambda = \frac{\Gamma(\alpha + r)}{\Gamma(\alpha) \cdot \Gamma(r)} \frac{\beta^\alpha T_\Sigma^r}{r(\beta + T_\Sigma)^{\alpha + r}}. \tag{3.83}$$

The first two moments of the number of failures are :

$$E_r = \sum_{r=0}^{\infty} r f(r) = \frac{\alpha T_\Sigma}{\beta}$$

$$E_{r^2} = \sum_{i=0}^{\infty} r^2 f(r) = \frac{\alpha T_\Sigma(\beta + T_\Sigma)}{\beta^2} + \left(\frac{\alpha T_\Sigma}{\beta}\right)^2. \tag{3.84}$$

The theoretical moments (3.84) are set equal to the empirical moments and the equations of the parameter estimates are obtained :

$$\bar{r} = \frac{\sum_{i=1}^{K} r_i}{K} = \frac{\alpha T_\Sigma}{\beta}$$

$$\bar{r}^2 = \frac{\sum_{i=1}^{K} r_i^2}{K} = \frac{\alpha T_\Sigma(\beta + T_\Sigma)}{\beta^2} + \left(\frac{\alpha T_\Sigma}{\beta}\right)^2. \tag{3.85}$$

Solving equations (3.85), we get the estimates of the prior parameters :

$$\hat{\beta} = \frac{\bar{r}\,T_\Sigma}{\overline{r^2} - (\bar{r})^2 - \bar{r}}\,; \quad \overline{r^2} - (\bar{r})^2 - \bar{r} > 0$$

$$\hat{\alpha} = \frac{\hat{\beta}\bar{r}}{T_\Sigma}\,. \tag{3.86}$$

There can be data sets for which the method of moments estimates is not meaningful. This occurs in estimating β in the prior distribution when $\bar{r} \geqslant \overline{r^2} - (\bar{r})^2$, causing the estimated β to be either infinite or a meaningless negative number. A modified method which provides for estimates of the prior parameter when $\bar{r} \geqslant \overline{r^2} - (\bar{r})^2$ is developed in [24]. The procedure is to place a prior lower bound on the coefficient of variation of the prior distribution. If the coefficient of variation as estimated by the method of moments is greater than the prior lower bound, the usual estimates (3.86) are computed. Otherwise, a modified estimate is defined as follows.

The coefficient of variation of the a-priori distribution is

$$\frac{\sigma_\lambda}{E_\lambda} = \frac{1}{\sqrt{\alpha}}\,. \tag{3.87}$$

The method of moments estimate of the coefficient of variation is, from the relations (3.86),

$$\frac{1}{\sqrt{\hat{\alpha}}} = \frac{\sqrt{\overline{r^2} - (\bar{r})^2 - \bar{r}}}{\bar{r}}\,. \tag{3.88}$$

Consider the case when the coefficient of variation is greater than a constant k. If the estimated coefficient of variation (3.88) is greater than k, the method of moments estimates remains unchanged. If it is less than or equal to k, or does not exist, α is estimated by setting the coefficient of variation equal to its prior lower bound and β is estimated by the method of moments. The modified estimates of α and β are then :

$$\hat{\alpha}^* = \hat{\alpha}\,; \quad \hat{\beta}^* = \hat{\beta} \quad \text{if } \hat{\alpha} < \frac{1}{k}$$

$$\hat{\alpha}^* = \frac{1}{k^2}\,; \quad \hat{\beta}^* = \frac{T_\Sigma}{k^2\bar{r}} \quad \text{if } \hat{\alpha} > \frac{1}{k^2} \tag{3.89}$$

where $\hat{\alpha}$ and $\hat{\beta}$ are the usual estimates given by relations (3.86).

The method uses the subjective prior information on the coefficient of variation together with the prior data, so that a combination of ordinary and empirical Bayes methods is achieved.

The modified method of moments estimates depends on a constant k which is the prior lower bound on the coefficient of variation (cv) of the prior distribution. The simulation performed in [24] points out that, for small samples, the method of moments estimates of the parameters of the prior are poor, but the modified method of moments substantially improves their mean squared errors. However, for larger sample sizes, the modified method of moments estimates provides only a small improvement, although the improvement is higher for larger values of α. The essential advantage of the modified method of moments estimates is that they always exist, whereas the basic method of moments estimates often do not exist, especially for small samples.

The empirical Bayes methods presented above describe the variability of the parameter to be estimated by an a-priori distribution of the gamma type and use the prior data to estimate the parameters of this a-priori distribution. It is possible to use the same data in order to test the hypothesis that the true a-priori distribution is actually of the gamma type. To this end, it is necessary to check the marginal distribution (3.75) or (3.83) against the data by performing a goodness-of-fit test. The power of such a test is low, as shown in [23], so the data which support the hypothesis of a gamma prior also support the hypothesis of a uniform prior. Thus although identification of the a-priori distribution is possible, the available methods are not sufficiently discriminating, and it is necessary to select the family of a-priori distributions using partially subjective reasons and then to specify an a-priori distribution by estimating its parameters from the prior data.

The objective interpretation of the a-priori distribution poses some procedural problems in the context of the empirical Bayes inference. Suppose that the a-priori distribution is specified on the basis of k samples, the k-th sample being the current experiment. The data provided by the current experiment are used not only in the estimation of the prior parameters but also to derive the a-posteriori distribution via the Bayes equation. For instance, in the methodology outlined above, the a-posteriori distribution is of the gamma type with parameters $\alpha + r_k$ and $\beta + T_{\Sigma_k}$, so that both a point and an interval estimate of the failure rate may be easily computed. When a new experiment (the $k + 1^{st}$) is performed, the usual Bayes procedure is to take as an a-priori distribution with respect to the new data the a-posteriori distribution derived after the experiment k. According to the objective interpretation of the a-priori distribution, such a procedure is not correct, as the λ value which corresponds to the experiment $k + 1$ is not generated by the a-posteriori distribution but by the a-priori distribution associated with the random parameter. The data provided by the $k + 1$ experiment must be used, together with the k previous samples, to estimate the parameters of the a-priori distribution and then the a-posteriori distribution with respect to the $k + 1$ sample must be derived according to the Bayes equation. The learning process which is a distinct feature of the bayesian inference is represented, in the context of empirical Bayes methods, by the continuous updating of the estimates of the prior parameters. When the a-priori distribution is subjectively interpreted, that is, the parameter of the reliability model is an unknown deterministic quantity and the a-priori distibution describes the uncertainty of the analyst concerning the true value of this parameter, the learning process consists of the continuous updating of the prior distribution via the Bayes equation, using as an a-priori distribution before an experiment the a-posteriori distribution resulting from the preceding experiment.

The difficulties implied by the parametric specification of the a-priori distribution led to another approach to empirical Bayes inference, which is non-parametric in the sense that the a-priori distribution is not analytically specified but is estimated directly from the prior data. Suppose that $k \geqslant 2$ prior estimates of the constant failure rate λ are available. If these estimates are ordered so that $\hat{\lambda}_1 \leqslant \hat{\lambda}_2 \leqslant \ldots \leqslant \hat{\lambda}_k$, the a-priori cdf may be non-parametrically estimated according to :

$$\hat{F}_a(\lambda) = \begin{cases} 0 & \lambda < \hat{\lambda}_1 \\ \dfrac{1}{k} & \hat{\lambda}_1 \leqslant \lambda < \hat{\lambda}_2 \\ \cdots\cdots & \\ \dfrac{i}{k} & \hat{\lambda}_i \leqslant \lambda < \hat{\lambda}_{i+1} \\ \dfrac{k-1}{k} & \hat{\lambda}_{k-1} \leqslant \lambda < \hat{\lambda}_k \\ 1 & \lambda \geqslant \hat{\lambda}_k . \end{cases} \tag{3.90}$$

The estimated a-priori cdf may be used, together with the data provided by the current experiment, to estimate the current value of the failure rate, which is the posterior mean :

$$\hat{\lambda} = \int_0^\infty \lambda f_p(\lambda)\, \mathrm{d}\lambda = \frac{\int_0^\infty \lambda f(x_k/\lambda)\, \mathrm{d}F_a(\lambda)}{\int_0^\infty f(x_k/\lambda) \mathrm{d}F_a(\lambda)} . \tag{3.91}$$

The empirical Bayes estimate (3.91) is obtained by replacing the unknown prior distribution by the step function (3.90), which has steps of equal height $1/k$ at each experiment

$$\hat{\lambda} = \frac{\sum_{i=1}^{k} \hat{\lambda}_i f(x_k/\hat{\lambda}_i)\, \mathrm{d}\hat{F}_a(\hat{\lambda}_i)}{\sum_{i=1}^{k} f(x_k/\hat{\lambda}_i)\, \mathrm{d}\hat{F}_a(\hat{\lambda}_i)} \tag{3.92}$$

where x_i are the data provided by the experiment i.

As $d\hat{F}_a(\hat{\lambda}_i) = 1/k$, $i = 1, 2, \ldots, k$, expression (3.92) may be written :

$$\hat{\lambda} = \frac{\sum_{i=1}^{k} \hat{\lambda}_i f(x_k/\hat{\lambda}_i)}{\sum_{i=1}^{k} f(x_k/\hat{\lambda}_i)}. \tag{3.93}$$

The estimate (3.93) is "smooth" in the sense that it is a function of all the information contained in the sequence of k experiments.

Taking now into account the actual form of $f(x_k/\hat{\lambda}_i)$ which was derived in Chapter 2, expression (3.93) becomes :

$$\hat{\lambda} = \frac{\sum_{i=1}^{k} \hat{\lambda}_i^{r_k+1} \exp[-\hat{\lambda}_i T_{\Sigma_k}]}{\sum_{i=1}^{k} \hat{\lambda}_i^{r_k} \exp[-\hat{\lambda}_i T_{\Sigma_k}]} \tag{3.94}$$

where (r_k, T_{Σ_k}) are respectively the numbers of failures and the total time on test observed in the current experiment.

If the estimates $\hat{\lambda}_i$ in (3.94) are obtained by the maximum likelihood principle, we have $\hat{\lambda}_i = r_i/T_{\Sigma_i}$ and the Bayes estimate (3.94) becomes :

$$\hat{\lambda} = \frac{\sum_{i=1}^{k} \left(\frac{r_i}{T_{\Sigma_i}}\right)^{r_k+1} \exp\left[-\frac{r_i T_{\Sigma_k}}{T_{\Sigma_i}}\right]}{\sum_{i=1}^{k} \left(\frac{r_i}{T_{\Sigma_i}}\right)^{r_k} \exp\left[-\frac{r_i T_{\Sigma_k}}{T_{\Sigma_i}}\right]}. \tag{3.95}$$

Any other reliability measure may be estimated if the relation between it and the parameter λ is known. For instance, the point estimate of the reliability function $R(t) = e^{-\lambda t}$ is obtained by taking the mean of $e^{-\lambda t}$ over the posterior distribution of λ :

$$\hat{R}(t) = \int_0^{\infty} e^{-\lambda t} f_p(\lambda)\, d\lambda = \frac{\int_0^{\infty} e^{-\lambda t} f(x_k/\lambda) dF_a(\lambda)}{\int_0^{\infty} f(x_k/\lambda) dF_a(\lambda)}. \tag{3.96}$$

Replacing the a-priori distribution by its non-parametric estimate (3.90), expression (3.96) becomes :

$$\hat{R}(t) = \frac{\sum_{i=1}^{k} e^{-\hat{\lambda}_i t} f(x_k/\hat{\lambda}_i)}{\sum_{i=1}^{k} f(x_k/\hat{\lambda}_i)}. \tag{3.97}$$

Taking into account the expressions of the maximum likelihood estimates of λ_i and the pdfs of the data x_k, we get:

$$\hat{R}(t) = \frac{\sum_{i=1}^{k} \left(\frac{r_i}{T_{\Sigma_i}}\right)^{r_k} \exp\left[-\frac{r_i(t + T_{\Sigma_k})}{T_{\Sigma_i}}\right]}{\sum_{i=1}^{k} \left(\frac{r_i}{T_{\Sigma_i}}\right)^{r_k} \exp\left[-\frac{r_i T_{\Sigma_k}}{T_{\Sigma_i}}\right]}. \tag{3.98}$$

The empirical Bayes estimators (3.94) or (3.98) should be superior to the usual estimators, which take into account only the current data. In order to prove this, the mean squared errors of the classical and empirical Bayes estimators were computed in [25] and [26] by Monte Carlo simulation. It was shown that the empirical Bayes estimators of λ of the reliability function have almost uniformly smaller mean squared errors than any of the conventional estimators. Moreover, the MSEs of the empirical Bayes estimators decrease as the number of experiments increases, but they are not a function of the prior distribution form. Most of the squared error improvement in the empirical Bayes concept over the conventional type is realized within 5 — 15 experiments. However, if an empirical Bayes procedure is to be of practical use in life testing, a significant squared error improvement over the conventional type must be realized with relatively few experiments. This is primarily because of the expected limited number of previous experiments. It is shown [25] that a significant improvement is achieved in that approximately 15 — 35% reduction in mean-squared error is realized with only five experiments.

The following chapters will further demonstrate the usefulness of the bayesian point of view for practical problems.

References

1. Carnap, R., The Object of Inductive Logic. Logic and Philosophy (in Romanian), Ed. politică, Bucharest, 1966.
2. Mihoc, Gh., Craiu, V., Treatise of Mathematical Statistics (in Romanian), vol. IV, Ed. Academiei, Bucharest, 1981.
3. de Groot, M., Optimal Statistical Decision (in Russian), Mir, Moscow, 1974.
4. Tsokos, C.P., Shimi, I.N. (eds). The Theory and Applications of Reliability with Emphasis on Bayesian and Nonparametric Methods, vols I—II, Academic Press, New York, 1977.
5. Beliş, M., The Mechanisms of Intelligence (in Romanian), Ed. tehnică, Bucharest, 1978.
6. Easterling, R., A personal view on the bayesian controversy in reliability and statistics, IEEE Trans. Reliab., R-21, 3(1972) 186—194.
7. Colombo, A.G., Constantini, D., Ground hypotheses for beta distribution as bayesian prior, IEEE Trans. Reliab., R-29, 1(1980) 17—21.
8. McFarland, W.J., Bayes' equation, reliability and multiple hypothesis testing, IEEE Trans. Reliab., R-21, 3 (1972) 136—147.
9. Cătuneanu, V.M., Mihalache, A., Reduction of Reliability Data Tests Cost by Use of a priori Information (in Romanian). The 1st Symposium on Electronic and Electrotechnic Products Reliability, ICPE, Bucharest, 1977.
10. Cătuneanu, V.M., Mihalache, A., Efficiency Analysis of Censored Reliability Tests (in Romanian), EEA Electrotehnica, 25, 3(1977) 101—104.

11. BARD, Y., Nonlinear Parameter Estimation, Academic Press, New York, 1974.
12. EVANS, R.A., Classical confidence intervals have a bayesian interpretation, Proceedings 1970 Annual Symposium on Reliability, 1970, 343—347.
13. CRELLIN, C.L., The philosophy and mathematics of Bayes' equation, IEEE Trans. Reliab., R-21, 3(1972) 131—135.
14. KASHYAP, R., Prior probability and uncertainty, IEEE, Trans. Inf. Theory, IT—17, 6(1971) 641—650.
15. WILSON, M.A., Experience with bayesian reliability measurement of large systems, IEEE Trans. Reliab., R-21, 3(1972) 181—185.
16. CANFIELD, R.V., TEED, J.C., Selecting the prior distribution in bayesian estimation, IEEE Trans. Reliab. R-26, 4(1977) 283—285.
17. MARITZ, J.S., Empirical Bayes Methods, Methuen, London, 1970.
18. MARTZ, H.F., Pooling life-test data by means of the empirical Bayes method, IEEE Trans. Reliab., R-24, 1(1975) 27—30.
19. BREIPOHL, A., Group reliability predictions, IEEE Trans. Reliab., R-27, 2(1978) 139—144.
20. BREIPOHL, A., Kalman filtering and its application to reliability, IEEE Trans. Reliab., R-18, 3(1969) 127—131.
21. HARRIS, C.M., SINGPURWALLA N.D., Life distributions derived from stochastic hazard functions, IEEE Trans. Reliab., R-17, 2(1968) 70—79.
22. SCHAFER, R.F., FEDUCCIA, A.J., Prior distribution fitted to observed reliability data, IEEE Trans. Reliab., R-21, 3(1972) 148—154.
23. HIGGINS, J.J., TSOKOS, C.P., On the behaviour of some quantities used in bayesian reliability demonstration tests, IEEE Trans. Reliab., R-25, 4(1976) 261—264.
24. HIGGINS J.J., TSOKOS, C.P., Modified method of moments in empirical Bayes estimation, IEEE Trans. Reliab., R-28, 1(1979) 27—31.
25. CANAVOS, G.C., An empirical Bayes approach for the Poisson life distribution, IEEE Trans. Reliab., R-22, 2(1973) 91—96.
26. LEMON, G.H., An empirical Bayes approach to reliability, IEEE Trans. Reliab., R-21, 3(1972) 115—158.
27. CANFIELD, R.V., A bayesian approach to reliability estimation using a loss function, IEEE Trans. Reliab., R-21, 3(1972) 70—74.
28. TILLMAN, F.A., KUO, W., HWANG, C.L., GROSH, D.L., Bayesian reliability and availability — a review, IEEE Trans. Reliab., R—31, 4(1982) 362—372.
29. TZE FEN LI, Empirical Bayes approach to reliability estimation of the exponential distribution, IEEE Trans. Reliab., R—33, 3(1984) 233—326.
30. COLOMBO, A.G., JAARSMA, R.J., Bayesian estimation of constant failure and unavailability, IEEE Trans. Reliab., R-31, 1(1983) 84—86.
31. PROHORENKO, V.A., GALIAKOV, V.F., Reliability Analysis Using a priori Information (in Russian), Nauka i Tehnika, Minsk, 1979.

Chapter 4
Reliability and stresses

The information required for accurate estimation of system reliability is obtained by carrying out long and costly life tests. The testing time and cost may be reduced by accelerated life testing, in which the units under test are run at higher than design stresses and fail sooner. The shortened liveg recorded in an accelerated test are then extrapolated, to estimate the reliability under design conditions. Such a procedure must be based on a model which describes the relation between the stress level and the reliability measure.

4.1 The stress-strength model

Interaction between a system and its environment appears when the system is subjected to stresses. Generally, the system has some influence on its environment, by perturbing other systems or by ecological implications, but this may be neglected for the moment.

The stresses applied to the system may be classified into useful and perturbing stresses. The useful stress is the load L on the system which accompanies the performance of its function. Such loads are the power dissipated in electrical and electronic components, the tensions in mechanical devices, etc. The perturbing stresses include temperature, pressure, vibration, humidity and all the stresses which are not essential to the functioning of the system. We shall denote them by T, as the temperature is probably the most important perturbing stress.

We have seen from the system theory approach to reliability that the stresses directly influence the component parameters, and only through this influence are the system performances affected.

The relation between stresses and component reliability may be expressed by the concept of strength. The strength S is the stress level which causes the component to fail. It is obvious that the strength is a random variable, because of the variability among the components in a batch. The component strength decreases with time until it becomes equal to the load applied, and then failure occurs. The reliability model (the time-to-failure distribution) may be derived from the initial distribution of the strength if the equation that governs the decrease of $S(t)$ is known.

This equation is based on the kinetics of chemical reactions, as the degradation of any material is due to such reactions. Suppose that the strength is a function of the concentration of the material, $S(t) = h[A(t)]$. The chemical reaction which is responsible for the degradation is

$$\mathrm{A} + B \rightarrow x \tag{4.1}$$

where $A(t)$ is the concentration in the material, $B(t)$ is the environmental concentration and $x(t)$ is the resulting concentration.

The Goldberg and Vaage law states that the variation of the resulting concentration is proportional to the concentrations of the reagents, A and B [1] :

$$\frac{\mathrm{d}x}{\mathrm{d}t} = K(A, B). \tag{4.2}$$

As the environmental concentration does not vary during the reaction, equation (4.2) may be written

$$\frac{\mathrm{d}x}{\mathrm{d}t} = KA(t) = K(A_0 - x(t)) \, ; \quad A_0 = A(0). \tag{4.3}$$

Integrating equation (4.3), we get :

$$\ln A(t) = \ln A_0 - Kt. \tag{4.4}$$

The constant K is the reaction rate and depends essentially on the temperature. According to the Arrhenius model, it has the form

$$K = C_1 e^{-\frac{Q}{kT}} \tag{4.5}$$

where k is the Boltzman constant, T is the absolute temperature and Q is the activation energy which is a constant typical of a specific material and a specific failure mechanism.

The reaction rate K may depend on other stresses than temperature. The Eyring model accounts for these stresses, giving for K the expression [2] :

$$K = C_1 T L^{C_2} \mathrm{e}^{-\frac{Q}{kT}} \tag{4.6}$$

where L is a stress, other than temperature.

The relationship $h(A)$ between the strength and the concentration may written in the form :

$$S(t) = C_3[A(t)]^{C_4} \tag{4.7}$$

with the condition :

$$S_0 = S(0) = C_3 A_0^{C_4}. \tag{4.8}$$

Taking the logarithms of equations (4.7) and (4.8) and using equation (4.4), the strength is given by

$$\ln S(t) = \ln S_0 - KC_4 t. \tag{4.9}$$

The constant C_4 may be included in the constant C_1 from expressions (4.5) or (4.6), so that equation (4.9) may be written :

$$\ln S(t) = \ln S_0 - Kt, \tag{4.10}$$

where the reaction rate K is given either by the Arrhenius model (4.5) or by the Eyring model (4.6).

Equation (4.10) states that the logarithm of the strength decreases linearly with time. The rate of degradation depends on the temperature and on the load applied. Let $g_0(s)$ denote the initial distribution of the strength. From Fig. 4.1 it is clear that the time-to-failure distribution may be derived from $g_0(s)$ and equation (4.10).

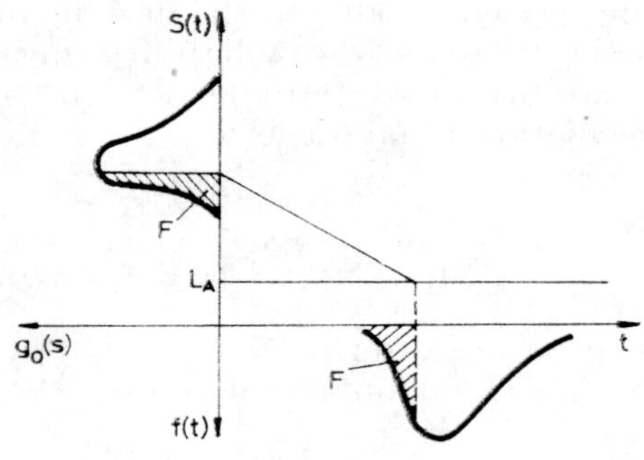

Fig. 4.1. — Derivation of the time-to-failure distribution in the stress-strength model.

The individual values of the strength decrease in parallel if the temperature is the same. Let us consider the F percentile of the strength S_F, that is F percent of the components have strength less than S_F. Equation (4.10) also describes the evolution of S_F :

$$\ln S_F(t) = \ln S_F(0) - Kt. \tag{4.11}$$

Failure occurs, with probability F, when $S_F(t)$ becomes equal to the load (the useful stress) actually applied, L_A. The F percentile of the time-to-failure is then :

$$t_F = \frac{1}{K} \ln \frac{S_F(0)}{L_A}. \tag{4.12}$$

Replacing the degradation rate K with the Arrhenius model (4.5) and taking the logarithm of relationship (4.12), we get an equation which relates the reliability measure t_F to the temperatuie stress T :

$$\ln t_F = \text{constant} + \frac{Q}{kT}. \tag{4.13}$$

The logarithm of the time until the failure occurs with probability F depends linearly on the inverse of the absolute temperature. Relation (4.13) is the Arrhenius law, as applied in reliability.

When the Eyring model (4.6) is used for the degradation rate K, equation (4.12) becomes :

$$t_F = \frac{1}{C_1 T} L^{-c_2} e^{\frac{Q}{kT}} \ln \frac{S_F(0)}{L_A}. \tag{4.14}$$

The load L in equation (4.14) represents stresses other than temperature and the load L_A the actual useful stress applied to the component.

Equations (4.13) and (4.14) may be used to derive the acceleration factors which allow extrapolation of the reliability measures from the test conditions to the design conditions. The acceleration factor is the ratio between the reliability measure in the accelerated test and the reliability measure under design conditions. In the Arrhenius model, the acceleration factor is :

$$\tau = \frac{t_F}{(t_F)_R} = \exp\left[\frac{Q}{k}\left(\frac{1}{T} - \frac{1}{T_R}\right)\right]. \tag{4.15}$$

where the subscript R defines the test conditions. In the Eyring model, the acceleration factor is :

$$\tau = \frac{t_F}{(t_F)_R} = \frac{T_R}{T} \cdot \left(\frac{L_R}{L}\right)^{c_2} \cdot \exp\left[\frac{Q}{k}\left(\frac{Q}{T} - \frac{1}{T_R}\right)\right]. \tag{4.16}$$

The acceleration factor may be used in a parametric approach to derive the relationship between the parameter values of the time-to-failure distribution for different stress conditions.

For instance, in the case of the exponential distribution the F percentile of the time-to-failure is :

$$t_F = \frac{1}{\lambda} \ln \frac{1}{1 - F}. \tag{4.17}$$

Consequently the ratio between the failure rates, under design and test conditions respectively is :

$$\frac{\lambda}{\lambda_R} = \left[\frac{t_F}{(t_F)_R}\right]^{-1} = \frac{1}{\tau}. \tag{4.18}$$

In the case of distributions which have two or more parameters, it is assumed that only the scale parameter changes with the stress applied. The conservation of the shape parameter under different stresses is explained by the fact that the overstress does not modify the failure mechanism. As a matter of fact, this is the main problem of accelerated testing. If the stresses are not high enough, few failures are observed. But the stresses must not be so high as to change the basic failure mechanism, otherwise any extrapolation rule becomes invalid.

If the time-to-failure is Weibull distributed, the F percentile of the time-to-failure is :

$$t_F = \left(\alpha \ln \frac{1}{1-F}\right)^{\frac{1}{\beta}}. \tag{4.19}$$

Assuming that the shape parameter β is independent of the stress level, the relation between the α values under design and test conditions is :

$$\frac{\alpha}{\alpha_R} = \left[\frac{t_F}{(t_F)_R}\right]^{\beta} = \tau^{\beta}. \tag{4.20}$$

Practical application of the extrapolation rules requires knowledge of the constants which appear in expressions (4.15) and (4.16) for the acceleration factors. The usual practice is to estimate the activation energy Q by regression. To do this, equation (4.11) is rewritten for a specific value of F, usually 50% :

$$\ln S_{0.5}(t) = \ln S_{0.5}(0) - Kt. \tag{4.21}$$

Expressing the degradation rate by the Arrhenius model, equation (4.21) becomes :

$$\ln S_{0.5}(t) = \ln S_{0.5}(0) - Kte^{-\frac{Q}{kT}}. \tag{4.22}$$

Relation (4.22) is plotted on a probability chart. The data plot assesses the validity of the model, that is, it checks relation (4.22) between strength, stress and exposure time, and also yields estimates of the model parameters.

To make a relationship plot, a semilog probability chart should be used so that the theoretical relationship would appear as a straight line. The median of the strength must be estimated, at least at three points, in order to fit a straight line between the points.

Such a procedure was employed in [3], using data on the dielectric breakdown strength of insulation specimens. The evolution of the dielectric strength at four test temperatures (180, 225, 250, 275°C) was studied during 64 weeks. The design temperature was 150°C and the failure criterion was 2 kV breakdown strength. The median of the initial strength was estimated as $\hat{S}_{0.5}(0) = 13.5$ kV. To estimate the constants K and Q (the activation energy), two widely separated absolute temperatures T_1 and T_2 and a long exposure time t were considered. From the fitted lines for T_1 and T_2, estimates $S^{(1)}_{0.5}(t)$ and $S^{(2)}_{0.5}(t)$ of the median breakdown strengths at t were obtained. From equations (4.22) written for $T = T_{1,2}$ we get :

$$\frac{Q}{k} = \left(\frac{1}{T_2} - \frac{1}{T_1}\right) \ln \frac{\ln \dfrac{S_{0.5}(0)}{S^{(1)}_{0.5}(t)}}{\ln \dfrac{S_{0.5}(0)}{S^{(2)}_{0.5}(t)}} ; \quad K = \frac{1}{t} e^{-\frac{Q}{kT_1}} \ln \frac{S_{0.5}(0)}{S^{(1)}_{0.5}(t)}. \tag{4.23}$$

Equations (4.23) may be solved after replacing the median strengths with their estimated values. For $T_1 = 250°C$, $T_2 = 275°C$, and $t = 32$ weeks, the plots give the estimates $\hat{S}_{0.5}(0) = 13.5$ kV, $\hat{S}_{0.5}^{(1)}(t) = 9.8$ kV and $\hat{S}_{0.5}^{(2)}(t) = 3.27$ kV.

From equations (4.23) we get :

$$\left(\frac{\hat{Q}}{k}\right) = 17065 \text{ K}; \quad \hat{K} = 6.375 \times 10^{11}/\text{week}. \tag{4.24}$$

Once the activation energy is estimated, the acceleration factor may be used to derive the reliability measures under design conditions from those estimated under test conditions.

The method described above uses data obtained in constant accelerated tests. A more general method in accelerated testing includes varying the stresses applied to the components. A stress-strength model adequate for the situation of a time-varying stress may be obtained by the accumulation principle [2]. When the stress is a time function, the degradation rate is also a time function, so that equation (4.10) of the strength becomes :

$$\ln S(t) = \ln S(0) - \int_0^t K(t)\mathrm{d}t. \tag{4.25}$$

In the case of a constant stress, $K(t) = K$ and equation (4.25) reduces to equation (4.10).

It is difficult in practice to achieve a continuously varying stress, so a step stress test is employed, that is, the components are subjected to the stress $T_1 < T_2 <, \ldots, T_n$ during the time-intervals $t_1, t_2, \ldots, t_n$. Equation (4.25) may be written in this case :

$$\ln S(t) = \ln S(0) - \sum_{i=1}^{n} K_i t_i \tag{4.26}$$

where K_i is the degradation rate at stress T_i. The time-to-failure X_i under the stress T_i may be obtained by equating the strength (4.10) to the stress actually applied L_A :

$$X_i = \frac{1}{K_i} \ln \frac{S(0)}{L_A}. \tag{4.27}$$

The time-to-failure under the successive stresses $T_1, T_2, \ldots, T_n$ is obtained by equating expression (4.26) to the load L_A. In this case, the time-to-failure is approximately $\sum_{i=1}^{n} t_i$, where n is the actual number of steps until the failure occurs :

$$\ln L_A = \ln S(0) - \sum_{i=1}^{n} K_i t_i. \tag{4.28}$$

From equations (4.27) and (4.28) we obtain the generalized Miner's equation [2]:

$$\sum_{i=1}^{n} \frac{t_i}{X_i} = 1. \tag{4.29}$$

An immediate application of Miner's equation is the following. Consider two stresses $T_1 < T_2$ and suppose that X_2, the shorter time-to-failure at the higher stress, is known. It is required to find X_1, the time-to-failure at the lower stress. From equation 4.29 we get :

$$X_1 = \frac{t_1 X_2}{X_2 - t_2} \tag{4.30}$$

where t_1 and t_2 are actual stress application durations at the lower stress and the higher stress, respectively.

The advantage of the method lies in the saving of testing information. Suppose that, at the intended stress, the test is truncated halfway because of an unexpectedly long test time or some other reasons. We can nevertheless calculate the lifetime of the item under the intended stress, by transferring these truncated samples into another stress condition in which the lifetime is already known. Miner's rule is also useful at the design stage in the prediction of component life when operating under various stress levels.

4.2 The parameter model

The influence of stresses on reliability may also be modelled on the basis of the time-varying technical parameters which describe a specific component. The life history of any particular device may be described by a critical parameter y, with initial value $y(0)$. Due to physical or chemical processes, the value of y changes as a function of time t, until a critical value $y_{\min}$ or $y_{\max}$ has been reached at which the device is beyond its specification. Without loss of generality it will be assumed that $y(t)$ is a decreasing function of time. It will also be assumed that failure occurs when $y(t)$ reaches a critical value $y_{\min} = \omega$, such that the time to failure X is the solution of the equation :

$$y(x) = \omega. \tag{4.31}$$

At some later time t_0, the parameter is practically equal to zero :

$$y(t_0) = 0. \tag{4.32}$$

Due to the manufacturing process the initial value $y_0 = y(0)$ is a random value, with the initial distribution $g_0(y)$. The physical/chemical processes which are responsible for the device degradation are governed by differential equations involving the time. It is assumed that the differential equation is of the first order. Since physical mechanisms are invariant with respect to time, the differential equation will not contain the time explicitly.

Suppose that the degeneration of the parameter y is determined by the differential equation :

$$\frac{dy}{dt} = -\frac{K(T)}{h(y)} \tag{4.33}$$

where the constant $K(T)$ depends strongly on temperature as, for instance, in the Arrhenius model.

Integration of equation (4.33) with the limit condition (4.32) and the initial condition $y(0) = y_0$ yields :

$$\int_0^{y(t)} h(y)\, dy = K(T)(t_0 - t). \tag{4.34}$$

The value of t_0 is found from equation (4.34) by putting $t = 0$:

$$t_0 = \frac{1}{K(T)} \int_0^{y_0} h(y)\, dy. \tag{4.35}$$

The time-to-failure x is obtained from equation (4.34), making $y(x)$ equal to ω :

$$\int_0^{\omega} h(y)\, dy = K(T)(t_0 - x). \tag{4.36}$$

It may be seen from equation (4.36) that the difference $\tau = t_0 - x$ is constant, irrespective of the initial value of the parameter, so that the time interval between the occurrence of a failure and the annulment of the parameter is constant. Thus the time-to-failure distribution may be derived from the distribution of t_0, using the simple translation $x = t_0 - \tau$.

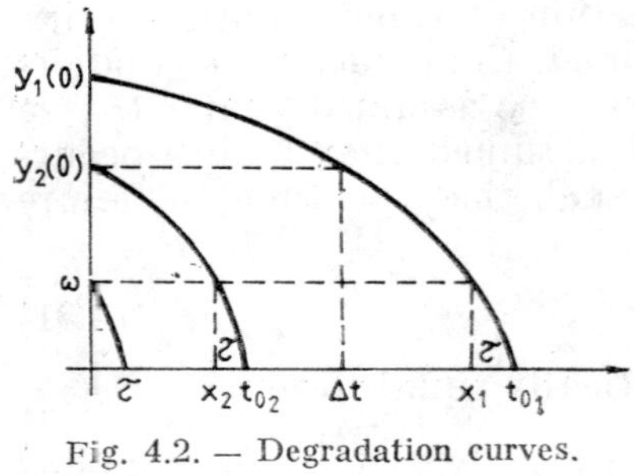

Fig. 4.2. — Degradation curves.

In Fig. 4.2, solutions of equation (4.34) are sketched for different values of y_0. The solutions corresponding to different constants of integration are obtained by a translation $\Delta t = t_{01} - t_{02}$ parallel to the time axis. The life history of a weak part $y_2(0)$ from $t = 0$ onwards is then the same as that of a strong part $y_1(0)$ from $t = \Delta t$ onwards, in which $y_1(\Delta t) = y_2(0)$. Therefore potential early failures are characterized, according to this model, by an infirmity of old age, rather than by an infantile complaint.

We shall derive the reliability model (the time-to-failure distribution) via the distribution of t_0. From equation (4.35), the distribution of t_0 is :

$$\varphi(t) = \frac{g_0(y)}{\left|\dfrac{dt_0}{dy_0}\right|} = \frac{g_0(y)K(T)}{h(y)} \tag{4.37}$$

where y is the random initial value of the parameter.

Once the distribution of t_0 is established, the time-to-failure distribution $f(x)$ may easily be derived from expression (4.37) by replacing t by $x + \tau$. The distribution (4.37) refers to the complete sample, including the devices with $y_0 < \omega$. If the latter have been removed from the sample by testing, the new distribution of t_0 is given by $\varphi(t + \tau)$.

A specific form of the function $h(y)$ may be assumed, that is :

$$h(y) = y^n. \tag{4.38}$$

For $n = 0$, the degradation rate is constant. For $n = -1$, $y(t)$ decreases exponentially with time. In general, $n \geqslant 0$. Taking into account expression (4.38), equation (4.35) becomes :

$$t_0 = \frac{y_0^{n+1}}{(n+1)\,K(T)} \tag{4.39}$$

The distribution of t_0 becomes, in this specific case,

$$\varphi(t) = \frac{g_0(y)y}{(n+1)t} \tag{4.40}$$

where y and t have been substituted for y_0 and t_0, respectively. Thus, the derivative of t_0 with respect to the initial value y_0 is :

$$\frac{dt_0}{dy_0} = \frac{y_0^n}{K(T)} = \frac{(n+1)t_0}{y_0}. \tag{4.41}$$

It is possible to derive some properties of the time-to-failure distribution from the characteristics of the initial distribution of the parameter values. Differentiation of equation (4.40) gives :

$$(n+1)\frac{d\varphi(t)}{dt} = -\frac{g_0(y)y}{t^2} + \frac{1}{t}\left[g_0(y) + y\frac{dg_0(y)}{dy}\right]\frac{dy}{dt}. \tag{4.42}$$

Taking account of expression (4.41), equation (4.42) becomes :

$$\frac{d\varphi(t)}{dt} = \frac{y}{(n+1)^2t^2}\left(-ng_0(y) + y\frac{dg_0(y)}{dy}\right). \tag{4.43}$$

If the variance of the manufacturing process is large, $g_0(y)$ is nearly constant and $\varphi(t)$ has a negative slope. If the precision of the manufacturing process is high, the derivative of $g_0(y)$ dominates and a positive slope of $\varphi(t)$ is to be expected.

To be specific, let us assume that y_0 is normally distributed with an average μ and a standard deviation σ, that is:

$$g_0(y) = \frac{\text{Constant}}{\sqrt{2\pi}\,\sigma} \mathrm{e}^{-\frac{1}{2}\left(\frac{y-\mu}{\sigma}\right)^2}. \tag{4.44}$$

Introducing expression (4.44) into equation (4.43), we get:

$$\frac{\mathrm{d}\varphi(t)}{\mathrm{d}t} = \frac{y}{\sqrt{2\pi}\,(n+1)^2 t^2}\left[-n + \frac{y(\mu - y)}{\sigma^2}\right] \mathrm{e}^{-\frac{1}{2}\left(\frac{y-\mu}{\sigma}\right)^2}. \tag{4.45}$$

If $\mu < 2\sigma\sqrt{n}$ the derivative (4.45) is always negative. This corresponds to the situation when the variance of the manufacturing process is large.

If $\mu > 2\sigma\sqrt{n}$, the slope of $\varphi(t)$ is negative for $y < \frac{1}{2}[\mu - \sqrt{\mu^2 - 4n\sigma^2}]$ or $y > \frac{1}{2}[\mu + \sqrt{\mu^2 - 4n\sigma^2}]$ and positive for $\frac{1}{2}[\mu - \sqrt{\mu^2 - 4n\sigma^2}] < y < \frac{1}{2}[\mu + \sqrt{\mu^2 - 4n\sigma^2}]$. This means that, for an accurate manufacturing process $\left(\sigma < \frac{\mu}{2\sqrt{n}}\right)$, the time-to-failure distribution has two extremum points.

The above considerations are valid for the complete distribution, from $y = 0$ (and $t = 0$) onwards. If part of the lot $y < \omega$ is rejected by testing, the pdf of the remaining batch refers to the region $t > \tau$ in which τ corresponds to ω. By testing, the curve is cut off at the left hand side. If $\omega < \frac{1}{2}[\mu - \sqrt{\mu^2 - 4n\sigma^2}]$, the pdf still has the two extremum points, while for $\frac{1}{2}[\mu - \sqrt{\mu^2 - 4n\sigma^2}] < \omega < \frac{1}{2}[\mu + \sqrt{\mu^2 - 4n\sigma^2}]$ it has only a maximum, and for $\omega > \frac{1}{2}[\mu + \sqrt{\mu^2 - 4n\sigma^2}]$ it is continuously decreasing, just as in the case $\mu < 2\sqrt{n}\sigma$.

The parameter model allowed derivation of the time-to-failure distribution from the initial distribution of the parameter and the degradation equation. The model shows that at a given instant the speed of degradation of the weak parts is greater than that of the strong parts, and their time-to-failure is disproportionately shorter. This "time lag" effect is particularly strong for small values of y_0, and may cause a decreasing time-to-failure distribution at the beginning. For large values of y, that is for larger t-values of the $f(t)$ curve, the time lag is less pronounced and when $\sigma \ll \mu$, the distribution will possibly rise. For $\sigma \gg \mu$ the parameter distribution $g_0(y)$ is flat, causing the time-to-failure distribution to decrease continuously. It may then be concluded that a low yield in general gives a high initial failure rate.

The parameter model which has been presented assumed that the degradation was governed by a differential equation, that is, the evolution of the parameter was deterministic and the random character of the time-to-failure was entirely due to the randomness of the parameter at $t = 0$. We shall now set this assumption aside and consider that the parameter $y(t)$ is a random non-stationary process [5]. The random function which describes the parameter degradation may be approximated by a linear function of the type:

$$y(t) = y_0 + st \tag{4.46}$$

where both the initial parameter y_0 and the slope s are random values.

Let $g_0(y)$ and $\psi(s)$ denote the distributions associated with y_0 and s respectively. Let (μ_0, σ_0) denote the mean and standard deviation of y_0 and (μ, σ) the mean and standard deviation of s. Finally, we shall denote by μ_y and σ_y the mean and standard deviation of $y(t)$.

From relation (4.46) it is easy to establish that:

$$\mu_y = \mu_0 + t\mu\,; \quad \sigma_y^2 = \sigma_0^2 + 2tK(y_0, s) + t^2\sigma^2 \tag{4.47}$$

where $K(y_0, s)$ is the correlation function between the initial value and the slope.

We shall assume that, on the basis of a substantial experimental effort, the values of μ_0, σ_0, μ, σ and $K(y_0, s)$ have been accurately estimated from the data.

The goal of the analysis is, as before, the derivation of the time-to-failure distribution (the reliability model). As the derivation is difficult in the general case, we shall consider two specific cases. Firstly, we shall assume that the variance of the slope σ^2 is sufficiently small that the slope s may be considered a deterministic value equal to its average μ. The parameter is a uniform random process and the time-to-failure X is

$$X = \frac{\omega - y_0}{\mu}. \tag{4.48}$$

The pdf $f(t)$ of the time-to-failure X is then

$$f(t) = \mu g_0(y). \tag{4.49}$$

From relation (4.49) it may be seen that, in the case of uniform random processes, the time-to-failure distribution is of the same form as the initial distribution of the parameter. For instance, if $g_0(y)$ is normal with parameters μ_0 and σ_0, $f(t)$ is also normal with parameters $\frac{\omega - \mu_0}{\mu}$ and $\frac{\sigma_0}{\mu}$:

$$f(t) = \mu \frac{1}{\sqrt{2\pi}\,\sigma_0} \exp\left[-\frac{1}{2}\left(\frac{\omega - \mu t - \mu_0}{\sigma_0}\right)^2\right]. \tag{4.50}$$

The relationship between the reliability measures and the stresses may easily be established, taking into account the fact that the slope μ is dependent on the stress (specifically on temperature, according to the Arrhenius model).

Let us now consider the situation when the variance σ_0 of the initial parameter is small, so that the initial value is deterministic and equal to μ_0. This applies for very accurate manufacturing processes or systems with adjustable parameters. In this last case it is assumed that, after each failure, the parameter is adjusted at the precise value μ_0 and drifts afterwards according to the equation $y(t) = \mu_0 + st$, until it reaches the failure limit ω. The time-to-failure is :

$$X = \frac{\omega - \mu_0}{s}. \tag{4.51}$$

The distribution of the time-to-failure is easily derived from the distribution $\psi(s)$ of the slope :

$$f(t) = \frac{\omega - \mu_0}{t^2}\,\psi(s). \tag{4.52}$$

To be specific, let $\psi(s)$ be a normal distribution with parameters μ and σ $(\mu > 3\sigma)$. The time-to-failure distribution will be :

$$f(t) = \frac{\beta}{\sqrt{2\pi}\,t^2}\exp\left[-\frac{1}{2}\left(\frac{\beta}{t} - \alpha\right)^2\right] \tag{4.53}$$

where : $\beta = \dfrac{\omega - \mu_0}{\sigma}$ and $\alpha = \dfrac{\mu}{\sigma}$.

The distribution (4.53) is known as the alpha distribution and is studied extensively in [5]. The relationship between the reliability measures and the stress is based on the dependence of the parameter α on the stress, mainly on temperature according to the Arrhenius model.

The parameter model allows derivation of the time-to-failure distribution from the initial distribution which describes the manufacturing process and the distribution of the degradation slope. The dependence of the parameters of the time-to-failure distribution on the stress is described by the same relationship as in the stress-strength model and is used in the practice of accelerated testing.

4.3 Accelerated life tests

Accelerated life testing is used to obtain information on the time-to-failure distribution more quickly. The units are run under severe conditions, and fail sooner than under normal conditions. A model is fitted to the accelerated failure times and then extrapolated to estimate the time-to-failure distribution under normal conditions. This is cheaper than testing under normal conditions, which is often impractical because of the duration of the test.

Accelerated test conditions involve higher than usual temperature, voltage, pressure, vibration, etc., or some combination of them. Constant stress is used because stress is constant in most applications and models based on constant stress are available (see Section 4.1). Tests at constant, high stresses can run too long because there is usually a great scatter in failure times. Step-stress testing is intended to reduce test time and to assure that failures occur quickly enough. A step-stress test runs through a pattern of specified stresses, each for a specified time, until the test specimen fails. The test stresses and their application times are chosen to ensure that failure occurs quickly enough. Often, different step-stress patterns are used on different specimens. Such tests can be further shortened by stopping the test with specimens not yet failed (censored data).

For a step-stress pattern, there is a specific distribution for time-to-failure under test. Data governed by this distribution are observed in the test. Not being interested in life under a step-stress pattern, one usually wants the life distribution under constant stress, which units experience in practical use. To analyse data, one needs a model that relates the distribution under step-stressing to the distribution under constant stress. Such a model is described in [6].

The model assumes that she remaining life of specimens depends only on the current cumulative fraction failed and current stress, regardless of how the fraction accumulated. Moreover, if held at the current stress, survivors will fail according to the distribution for that stress but starting at the previously accumulated fraction failed.

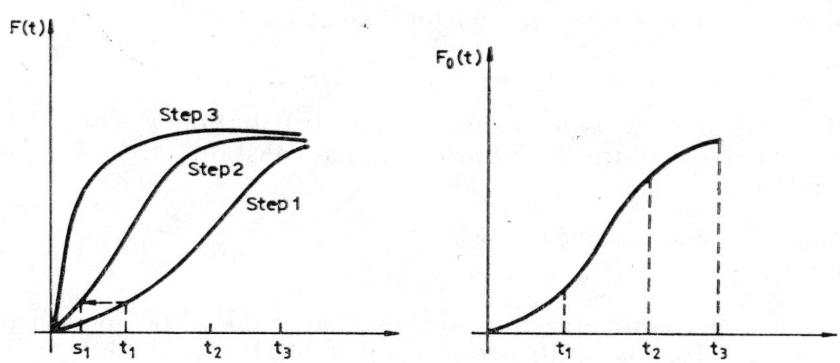

Fig. 4.3. — Relationship between constant and step-stress distributions.

Figure 4.3 depicts three cdfs for the constant stresses X_1, X_2, X_3, which may be temperature or voltage stresses. The arrows show that the specimens first follow the cdf for X_1 up to the first hold time t_1. When the stress increases from X_1 to X_2, the unfailed specimens continue along the cdf for X_2, starting at the accumulated fraction failed. The cdf $F_0(t)$ for the time-to-failure under the step-stress pattern consists of the segments of the cdfs for the constant stresses. Suppose that, for a particular pattern, step i runs at stress X_i, starts at time t_{i-1} and runs to time t_i ($t_0 = 0$). Denoting by $F_i(t)$ the cdf of the time-to-failure under a constant stress X_i, the fraction of specimens failing in step s is:

$$F_0(t) = F_1(t)\,; \quad 0 < t \leqslant t_1. \tag{4.54}$$

Step 2 has an equivalent start time s_1, which would have produced the same failed fraction under stress X_2, so that it is the solution of the equation:

$$F_2(s_1) = F_1(t_1). \tag{4.55}$$

The cumulative fraction of specimens failing in step 2 by total time t is:

$$F_0(t) = F_2(t - t_1 + s_1)\,; \quad t_1 \leqslant t < t_2. \tag{4.56}$$

Similarly, step 3 has the equivalent start time s_2 that is the solution of

$$F_3(s_2) = F_2(t_2 - t_1 + s_1). \tag{4.57}$$

The cumulative fraction of specimens failing in step 3 by total time t is:

$$F_0(t) = F_3(t - t_2 + s_2) \quad t_2 \leqslant t < t_3. \tag{4.58}$$

In general, step i has the equivalent start time s_{i-1} that is the solution of:

$$F_i(s_{i-1}) = F_{i-1}(t_{i-1} - t_{i-2} + s_{i-2}). \tag{4.59}$$

The cdf under the step-stress pattern is given by:

$$F_0(t) = F_i(t - t_{i-1} + s_{i-1}) \quad t_{i-1} \leqslant t < t_i. \tag{4.60}$$

We shall derive, as an example, the distribution $F_0(t)$ in the case when the time-of-failure distribution under constant stress is Weibull, that is:

$$F_i(t) = 1 - e^{-\frac{t^\beta}{\alpha_i}} \tag{4.61}$$

where the shape parameter β is constant and the scale parameter α_i depends on the stress according to relation (4.20). It is difficult to compute $F_0(t)$ when both temperature and useful stresses are considered, so the analysis will be restricted to the case of useful stresses L_i (e.g. voltage stresses for capacitors). For $T = T_R$, relations (4.16) and (4.20) give the dependence of α on the stress L:

$$\left(\frac{\alpha}{\alpha_R}\right)^{\frac{1}{\beta}} = \left(\frac{L_R}{L}\right)^{C_2}. \tag{4.62}$$

With a convenient notation, relation (4.62) becomes :

$$\alpha_i^{\frac{1}{\beta}} = c_1\left(\frac{L_R}{L_i}\right)^{c_2}. \tag{4.63}$$

The time-to-failure distribution under constant stress is then :

$$F_i(t) = 1 - \exp\left\{-\left[\frac{t}{c_1}\left(\frac{L_i}{L_R}\right)^{c_2}\right]^\beta\right\}. \tag{4.64}$$

The time-to-failure distribution for step 1 under the step stress pattern is :

$$F_0(t) = 1 - \exp\left\{-\left[\frac{t}{c_1}\left(\frac{L_1}{L_R}\right)^{c_2}\right]^\beta\right\}; \quad 0 < t \leqslant t_1. \tag{4.65}$$

The equivalent time s_1 at L_2 is given by (4.55) as :

$$s_1 = t_1\left(\frac{L_1}{L_2}\right)^{c_2}. \tag{4.66}$$

For step 2, we have :

$$F_0(t) = 1 - \exp\left\{-\left[\frac{t - t_1 + s_1}{c_1}\left(\frac{L_2}{L_R}\right)^{c_2}\right]^\beta\right\}; \quad t_1 \leqslant t < t_2. \tag{4.67}$$

The equivalent time s_2 is :

$$s_2 = (t_2 - t_1 + s_1)\left(\frac{L_2}{L_3}\right)^{c_2}. \tag{4.68}$$

In general, for step i :

$$s_{i-1} = (t_{i-1} - t_{i-2} + s_{i-2})\left(\frac{L_{i-1}}{L_i}\right)^{c_2};$$

$$F_0(t) = 1 - \exp\left\{-\left[\frac{t - t_{i-1} + s_{i-1}}{c_1}\left(\frac{L_i}{L_R}\right)^{c_2}\right]^\beta\right\}; \quad t_{i-1} \leqslant t < t_i. \tag{4.69}$$

Thus, the time-to-failure distribution under the step-stress pattern consists of segments of Weibull distributions.

The data actually observed are governed by the distribution (4.69). The parameters c_1, c_2, L_R and β of $F_0(t)$ are estimated from the data by the maximum likelihood method. Suppose that each specimen is subjected to a specific step stress pattern. Then each time-to-failure is distributed according to a specific $F_0(t)$. The distributions for different specimens have different t_i, s_i and L_i values, but the same parameters c_1, c_2, L_R and β. The pdf of the time-to-failure of each specimen is the time derivative of $F_0(t)$. The sample likelihood is then the product of such pdfs evaluated at failure times and of such survival functions $(1 - F_0(t))$ evaluated at censoring times. This sample likelihood is used to obtain ML point estimates and confidence limits for the parameters c_1, c_2, L_R, β. Once the parameters are specified, the reliability model is completed, as it describes the probability of failure in any time interval as a function of the stress.

We shall now consider the case of a simple step-stress test [7] when only two stress levels (X_1 and X_2) are used. In a time-step test, units are initially placed on test at stress X_1 and run until time t_1, when the stress is changed to X_2 and the test is continued until all units run to failure. In a failure-step test, the stress is changed to X_2 when a fraction P_1 of the units has failed at stress X_1, and the remaining fractions are then run to failure.

The time-to-failure under constant stress is assumed to be exponentially distributed, with parameter $\lambda(X)$. The functional relationship between the failure rate λ and the stress is easily derived from equation (4.18) and the expressions of the acceleration factor (4.15) or (4.16).

As it is analytically difficult to consider the useful and the perturbing stresses simultaneously, they will be treated separately, using for the acceleration factor the Arrhenius model (equation (4.15)) when temperature stress is considered and the Eyring model for $T = T_R$ (equation (4.16)) when the useful stress (e.g. voltage) is considered. The expressions for the failure rate are respectively :

$$\frac{\lambda}{\lambda_R} = \exp\left[-\frac{Q}{R}\left(\frac{1}{T} - \frac{1}{T_R}\right)\right]; \quad \frac{\lambda}{\lambda_R} = \left(\frac{L}{L_R}\right)^{c_2}. \tag{4.70}$$

Both equations (4.70) may be written in a unified manner as :

$$\ln\lambda = \gamma_1 + \gamma_2 X \tag{4.71}$$

where γ_1 and γ_2 are constants and X is either the reciprocal of absolute temperature or the logarithm of the useful stress.

Using expression (4.71), the time-to-failure distribution is :

$$F_i(t) = 1 - e^{-\lambda_i t} = 1 - e^{-te^{\gamma_1 + \gamma_2 x_i}}. \tag{4.72}$$

The time-to-failure distribution under the simple step-stress pattern is, according to relations (4.58) and (4.59),

$$F_0(t) = \begin{cases} 1 - e^{-\lambda_1 t} & 0 \leqslant t < t_1 \\ 1 - e^{-\lambda_2(t - t_1 + s_1)} & t \geqslant t_1 \end{cases} \tag{4.73}$$

where

$$s_1 = t_1 \frac{\lambda_1}{\lambda_2}.$$

The parameters γ_1 and γ_2 of $F_0(t)$ are estimated by the maximum likelihood principle, using the data observed in the simple step-stress test [7]. Then the failure rate at the design stress may be estimated according to equation (4.71).

It is possible to optimize the simple step-stress plan by minimizing the asymptotic variance of the ML estimator of the failure rate at a specified design stress X_D. In the case of the time step test this variance is minimized by the optimum choice of the duration t_1 of the first step, while in the case of the failure-step test, the variance is minimized by appropriately choosing the specified fraction P_1 failing at the first step X_1, at which point the stress is switched to X_2.

The optimization is performed in [7]. In a time-step test, the optimal duration of the first step which minimizes the variance of $\hat{\lambda}(X_D)$ is :

$$t_1^* = \frac{1}{\lambda_1} \ln \frac{X_1 + X_2 - 2X_D}{X_1 - X_D} \tag{4.74}$$

where X_D is the design stress.

It is obvious that relationship (4.74) holds whether the higher or the lower stress is applied first. If the lower stress is close to the design stress, t_1^* tends to infinity when the lower stress is applied first and to zero when the higher stress is applied first. This means that the units should be run only at the lower stress. If both stresses are much higher than the design stress, t_1^* tends to $(\ln 2)/\lambda_1$, which is the median of the time-to-failure at stress X_1 (about half of the units should fail at each stress).

In a failure-step test, the optimum failed fraction at the first step which minimizes the variance of $\hat{\lambda}(X_D)$ is :

$$P^* = \frac{X_2 - X_D}{X_1 + X_2 - 2X_D}. \tag{4.75}$$

If the lower stress is close to the design stress, P^* tends to unity when the lower stress is applied first and to zero when the higher stress is applied first. This means that the units should be run only at the lower stress. If both stresses are much higher than the design stress, P^* tends to 1/2, which means that half of the units should fail at each stress. In general, $1/2 < P^* < 1$ when the lower stress is applied first and $P^* < 1/2$ when the higher stress is applied first. Always, the majority of units should fail at the lower stress.

A difficulty of the optimization method is that the optimal duration t_1^* in a time step test depends on the unknown value of the parameter λ. A reasonable guess of the λ value at the first stress (either high or low) permits the design of a sub-optimal time step stress. However, the variance of the failure rate estimate will be larger than expected.

The approach to accelerated testing outlined above involves making two kinds of assumptions. Firstly, it is assumed that at any stress level, the time-to-failure distribution is a member of the same parametric family of distributions, such as Weibull, as in the case discussed above.

Secondly, a functional relationship is assumed between the parameters of the time-to-failure distribution and the applied stress. Such a functional relationship (the one given by relation (4.63), for instance) is derived from the acceleration factor τ which may be of the Arrhenius or of the Eyring type (relations (4.15) and (4.16)).

In many practical situations the different stress levels can have different effects on the failure mechanisms and, from an engineering point of view, it is therefore more realistic to allow for different forms of the time-to-failure distribution at different stress levels. Moreover, the acceleration factor was derived from partly empirical considerations, so that its validity should not be exaggerated.

For such reasons, a method has been proposed [8] to make inferences from accelerated life tests without making any distributional assumptions and without using any functional relationship for the parameters. Let $X_1 > X_2 > \ldots > X_k$ be the accelerated stresses and $X_D < X_k$ the design stress. If $z_i(t)$ is the time-dependent failure rate under stress X_i, it is reasonable to assume that

$$z_1(t) > z_2(t) > \ldots > z_k(t) > z_D(t). \tag{4.76}$$

Using failure data observed under stresses $X_1, X_2, \ldots, X_k$, estimates $z_j(t)$ of the failure rates $z_j(t)$ for $j = 1, 2, \ldots, k$ are obtained, such that for any t:

$$\hat{z}_1(t) \overset{st}{\geqslant} \hat{z}_2(t) \overset{st}{\geqslant} \hat{z}_3(t) \overset{st}{\geqslant} \ldots \overset{st}{\geqslant} \hat{z}_k(t). \tag{4.77}$$

The notation $X \overset{st}{\geqslant} Y$ denotes that X is stochastically larger than Y, that is

$$p(X \geqslant t) \geqslant p(Y \geqslant t) \qquad \text{for all } t. \tag{4.78}$$

As no assumption was made on the time-to-failure distribution, the time-dependent failure rate $z_j(t)$ will be approximated by a piecewise constant function. Let $(0, T)$ be the time interval of interest.

The time interval $(0, T)$ is divided into m sub-intervals of length h, so that $mh = T$. Thus, the piecewise constant function which approximates the true failure rate is :

$$z_j(t) = z_j(t_i) = \lambda_{ji}\,; \quad t_i \leqslant t < t_i + h \tag{4.79}$$

where $t_i = (i-1)h\,; \quad i = 1, 2, \ldots, m\,; \quad j = 1, 2, \ldots, k$.

The values $F_{ji} = \lambda_{ji}h$ represent the conditional probabilities of an item failing in the time interval $(t_i, t_i + h)$ under stress X_j, given that it has survived up to t_i. For h sufficiently small, F_{ji} is close to the probability of failure in (t_i, t_{i+h}).

According to the method described in [10], prior distributions are assigned to F_{ji} which describe the degree of belief of the analyst before any experiment is performed. It is convenient to use the family of beta distributions with parameters α and β_j :

$$f_a(F_{ji}) = \frac{\Gamma(\alpha + \beta_j)}{\Gamma(\alpha)\Gamma(\beta_j)} F_{ji}^{\alpha-1}(1 - F_{ji})^{\beta_j - 1}. \tag{4.80}$$

The stochastically monotonic condition of expression (4.77) implies that the a-priori means of F_{ji} should be decreasing with the stress. As the mean of a beta distribution is $\alpha/(\alpha + \beta_j)$, it follows that β_j must be a non-decreasing function of j, that is

$$\beta_j \geqslant \beta_{j-1}; \quad j = 2, 3, \ldots, k. \tag{4.81}$$

Thus the prior parameter β_j indicates the relative severity of the stress X_j.

Let n_{ji} be the number of items on test at time t_i under stress X_j, and r_{ji} the number of failures observed in $(t_i, t_i + h)$. The distribution of r_{ji} conditioned by F_{ji} is of the binomial type:

$$P(r_{ji}/F_{ji}) = C_{n_{ji}}^{r_{ji}} F_{ji}^{r_{ji}} (1 - F_{ji})^{n_{ji} - r_{ji}}. \tag{4.82}$$

According to the methods of bayesian estimation described in Chapter 3, the prior belief is combined with the experimental information via the Bayes equation into the a-posteriori distribution $f_p(F_{ji})$. As the beta distribution is the natural conjugate of the binomial distribution, the a-posteriori distribution is also of the beta type, with parameters $\alpha + r_{ji}$, $\beta_j + n_{ji} - r_{ji}$. As $n_{j,i+1} = n_{ji} - r_{ji}$, we may write:

$$f_p(F_{ji}) = \frac{\Gamma(\alpha + \beta_j + n_{ji})}{\Gamma(\alpha + r_{ji})\,\Gamma(\beta_j + n_{j,i+1})} F_{ji}^{\alpha + r_{ji} - 1} (1 - F_{ji})^{\beta_j + n_{j,i+1} - 1}. \tag{4.83}$$

The point estimator of F_{ji} is the a-posteriori mean:

$$\hat{F}_{ji} = \frac{\alpha + r_{ji}}{\alpha + \beta_j + n_{ji}}. \tag{4.84}$$

The a-posteriori means (4.84) should also be decreasing with j. To be assured that this condition is satisfied, it is sufficient that for every fixed value of i:

$$r_{ji} \leqslant r_{j-1,i} \tag{4.85a}$$

$$\beta_j + n_{ji} - r_{ji} \geqslant \beta_{j-1} + n_{j-1,i} - r_{j-1,i} \tag{4.85b}$$

$$j = 2, 3, \ldots k.$$

The conditions (4.85) state that the number of failures in interval i under X_j must not be greater than the number of failures in the same interval under X_{j-1}, for all values of j. A reasonable strategy is to have more items initially on test under the more severe stress X_{j-1} than under the stress X_j, so that $n_{j1} < n_{j-1,1}$ for all j.

Since $n_{ji} = n_{j,i-1} - r_{j,i-1}$, the conditions (4.85b) can be written as:

$$\beta_j + n_{j,i+1} \geqslant \beta_{j-1} + n_{j-1,i+1}; \quad j = 2, 3, \ldots, k. \tag{4.86}$$

If the conditions (4.85) are not satisfied by the random values of r_{ji} and n_{ji}, a pooling procedure for the adjacent violators must be used [10].

Consider the time interval i; by the conditions (4.86), it is required that:

$$r_{1,i} \geqslant r_{2,i} \geqslant \ldots \geqslant r_{j-1,i} \geqslant r_{ji} \geqslant \ldots \geqslant r_{ki};$$

$$\beta_1 + n_{1,i+1} \leqslant \beta_2 + n_{2,i+1} \leqslant \ldots \leqslant \beta_{j-1} + n_{j-1,i+1} \leqslant$$

$$\leqslant \beta_j + n_{j,i+1} \leqslant \ldots \leqslant \beta_k + n_{k,i+1}. \tag{4.87}$$

If a reversal occurs, that is, if either

$$r_{j-1,i} < r_{ji} \tag{4.88}$$

or if $\beta_{j-1} + n_{j-1,i} - r_{j-1,i} > \beta_j + n_{j,i} - r_{ji}$ then the violators must be pooled and replaced as shown below [10].

Both $r_{j-1,i}$ and r_{ji} are replaced by half their sum and both $\beta_{j-1} + n_{j-1,i} - r_{j-1,i}$ and $\beta_j + n_{ji} - r_{ji}$ by half their sum.

The resulting sequences are tested to see if they are properly ordered. If not, the adjacent violators are replaced by their appropriate averages, as before. The above procedure is followed for all time intervals and continued until all reversals are eliminated. Excessive pooling will occur if the stresses are too similar to each other. The point estimate of F_{ji} is computed according to relation (4.84), where r_{ji} and $\beta_j + n_{ji}$ are replaced by their pooled values.

The estimates $\hat{F}_{ji}$ are then used to obtain $\hat{F}_{Di}$, the probability of failure in the time interval i under the design stress.

One may use a functional relationship of the Arrhenius or Eyring type and estimate its parameters from $\hat{F}_{ji}$, $j = 1, 2, \ldots, k$ and then extrapolate the relationship to the design stress.

In the absence of any physical or engineering knowledge about the relationship between the probability of failure and the corresponding stress, the following simple, reasonable relations among F_{ij} may be used:

$$F_{ki} = w_0 + w_1 F_{k-1,i} + \ldots + w_{k-1} F_{1,i} \tag{4.89}$$

$$i = 1, 2, \ldots, m.$$

Relationship (4.89) states that the probability of failure over a particular time interval under stress X_{kj} is a linear function of the probabilities of failure under stresses $X_{k-1}, X_{k-2}, \ldots, X_1$.

Let $\hat{w}_0, \hat{w}_1, \ldots, \hat{w}_{k-1}$ be the least squares estimators of $w_0, w_1, \ldots, w_{k-1}$. These can be obtained from the relationship (4.89), using the estimates (4.84) of the failure probabilities.

Thus, for $i = 1, 2, \ldots, m$, we have

$$\hat{F}_{Di} = \hat{w}_0 + \hat{w}_1 \hat{F}_{ki} + \hat{w}_2 \hat{F}_{k-1,i} + \ldots + \hat{w}_{k-1} \hat{F}_{2,i} \tag{4.90}$$

as estimators of the probability of failure under design stress X_D. Since a unit must ultimately fail, we also have:

$$\hat{F}_{D,m+1} = 1 - \sum_{i=1}^{m} \hat{F}_{Di} \tag{4.91}$$

where $m + 1$ refers to the final time interval $(T, + \infty)$.

The reliability function up to time t_{r+1} under the design stress is:

$$\hat{R}_D(t_{r+1}) = \prod_{i=1}^{r} (1 - \hat{F}_{Di}). \tag{4.92}$$

The stress-strength and the parameter models describe the relationship between reliability and stresses. These models and the associated practical methods provide a detailed description of the fundamental state equation

$$\dot{X} = A(X, U) \tag{4.93}$$

where U is the stress vector and $\dot{X}$ the component parameters. This equation was derived in the first Chapter on the basis of system theory principles and it governs at the general level all degradation equations described in this Chapter.

References

1. Haviland, R.C., Techniques de Fiabilité et Durée de Vie des Equipements, Eyrolles, Paris, 1966.
2. Shiomi, H., Application of cumulative degradation model to accelerated life test, IEEE Trans. Reliab., R-17, 1 (1968) 27—33.
3. Nelson, W., Analysis of performance-degradation data from accelerated tests. IEEE Trans. Reliab., R-30, 2 (1981) 149—155.
4. Bosch, G., Model for failure rate curves, Microelectron. Reliab., vol. 19, 4 (1979) 371—375.
5. Drujinin, G.V., System Reliability (in Romanian), Ed. tehnică, Bucharest, 1968.
6. Nelson, W., Accelerated life testing — step-stress models and data analyses, IEEE Trans. Reliab., R-29, 2 (1980) 103—108.
7. Kielpinski, T.J., Nelson, W., Optimum censored accelerated tests for normal and lognormal life distributions, IEEE Trans. Reliab., R-24, 5(1975) 310—320.
8. Meeker, W.Q., Nelson, W., Optimum accelerated life-tests for the Weibull and extreme value distributions, IEEE Trans. Reliab., R-25, 5 (1975) 321—332.
9. Miller, R., Nelson, W., Optimum simple step-stress plans for accelerated life testing, IEEE Trans. Reliab., R-32, 1(1983) 59—65.
10. Proschan, F., Singpurwalla, N.D., A new approach to inference from accelerated life tests, IEEE Trans. Reliab., R-29, 2 (1980) 98—102.
11. Herr, E.A., Poe, A., Fox, A., Reliability evaluation and production for discrete semiconductors, IEEE Trans. Reliab., R-29, 3 (1980) 208—216.
12. Reich, B., A study of accelerated storage test conditions applicable to semiconductor devices and microcircuits, IEEE Trans. Reliab., R-27, 3 (1978) 178—180.
13. Rickers, H.C., Manno, P.F., Microprocessor and LSI microcircuit reliability — prediction model, IEEE Trans. Reliab., R-29, 3 (1980) 196—202.

Chapter 5

System renewal

5.1 Renewal processes

In the previous chapters, reliability analysis was restricted to the prediction of the first system failure. The models developed there described the degradation of system performance without taking into account the external actions which counteract this degradation, by partial or total reconditioning of the system. The evolution of a large class of systems naturally ends when the first failure occurs, because external intervention is not practicable. This is the class of non-repairable systems. For other systems, generally of high complexity, renewal possibilities exist, and their effectiveness therefore depends not only on their intrinsic reliability but also on the characteristics of the maintenance actions.

In order to take the maintenance actions into account, the concept of reliability must be re-interpreted as an overall capacity of the system to accomplish a specified task. The traditional limitation of the reliability concept, to the study of the system behaviour up to the first failure, has promoted the term effectiveness to describe the reliability of maintained systems [1]. We shall favour the term reliability in the broad sense mentioned above.

Reliability in the broad sense, or system effectiveness, is achieved by a combination of three factors. The first is the intrinsic reliability, which is the property of continued unfailing performance. The second, maintainability, is the capacity to restore the performance, and depends on the way the system is designed so that fault detection and correction can be easily achieved. Thirdly, the characteristics of the maintenance procedures influence reliability by the planning of spare parts and actions to inhibit the system wearing out.

The reliability model incorporating the maintenance actions may be developed at a global or at a structural level. In this Chapter, a global model which describes system reliability in the broad sense will be derived. The structural models will be discussed in Chapter 6.

We shall consider for the moment only those corrective maintenance actions (repairs) which are performed when the system fails.

We shall assume that a system failure can be easily detected, and that the duration of the repair is negligible. We shall denote by a renewal any corrective action (repair), irrespective of the impact it may have on the system performance. The evaluation of the system may then be viewed as a succession of renewals occurring at times $t_1, t_2, \ldots, t_n$ such that $X_i = t_i - t_{i-1}$, $i = 1, 2, \ldots, n, \ldots, t_0 = 0$.

Consider the time interval $(0, t)$. The number of renewals N_t occurring in this interval is a discrete stochastic process, called a renewal process. Once the characteristics of this process are known the reliability model is complete, as predictions of the evolution of the system can be made.

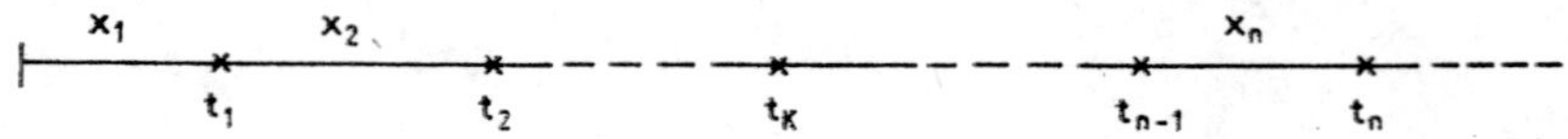

Fig. 5.1. — Renewal process.

Knowledge of the stochastic renewal process implies the possibility of computing the probabilities $P(N_t = r)$ for any t and r or, at least, of specifying some numerical characteristics of N_t like the mean and the variance. In order to do this, some assumptions about the renewals are made, which refer to both the characteristics of the system and of the maintenance actions.

A general assumption is that the stochastic renewal process is homogeneous, that is, the probability of more than one renewal occurring in the time interval $(t, t + \Delta t)$ is negligible. More precisely, the probability of several renewals occurring in $(t, t + \Delta t)$ is a function $\varepsilon(\Delta t)$ which tends to zero when $\Delta t \to 0$ more rapidly than Δt.

Another general assumption concerns the independence of the random time intervals $X_1, X_2, \ldots, X_n, \ldots$ which means that the behaviour of the system after any renewal is independent of its behaviour before the renewal. This is nearly always true, as in the most general case the system changes completely by renewal, as far as reliability is concerned. If this is the case, it is unnecessary to describe the system synthetically by the renewal process N_t, since it is more natural to describe each time interval X_i by the general reliability measures developed in Chapter 2. Thus the system is differently characterized in each time interval, without considering any global measure of reliability.

However, the system does not entirely change by renewal. The renewal may improve the reliability characteristics or may degrade them, depending on the maintenance actions.

The renewals may be classified on the basis of the reliability function $R_i(x)$ associated with the time interval X_i.

A proper renewal (or, shortly, a renewal) is the maintenance action which brings the system to the same state, eliminating the wear accumulated since the preceding renewal. For proper renewals, we have :

$$R_1(x) = R_2(x) = \ldots = R_n(x) = \ldots = R(x), \tag{5.1}$$

and the renewal process is simple.

The proper renewals modify the failure rate of the system by introducing jumps at the instants of failure. Such jumps are negative in the case of IFR systems, as shown in Fig. 5.2 (a).

If the renewals bring the system to the same state, which is different from the state at $t = 0$, the relation (5.1) becomes :

$$\begin{aligned} &R_2(x) = R_3(x) = \ldots = R(x) \\ &R_1(x) \neq R(x) \end{aligned} \qquad (5.2)$$

and N_t is called a general renewal process.

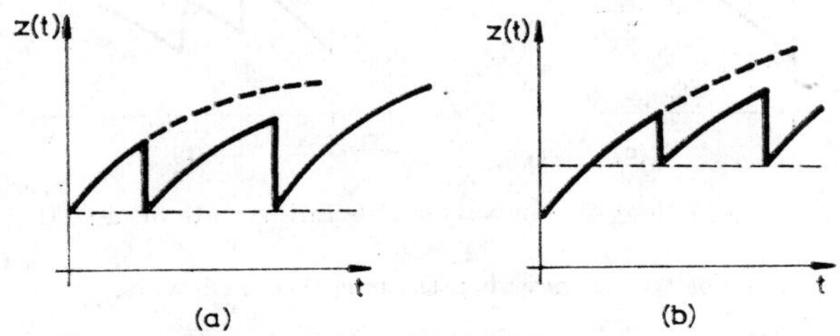

Fig. 5.2. — Effect of proper renewals on the failure rate of an IFR system :
(a) simple renewal process ; general renewal process.

In a simple renewal process $X_1, X_2, \ldots, X_n \ldots$ are iid random variables, and the reliability measures are the same for all time intervals between renewals. The MTTF (mean time to failure) which is the mean of X_i is the same for all i, so that it can be interpreted as an MTBF (mean time between failures).

In a general renewal process, $X_2, X_3, \ldots, X_n$ are iid random variables while X_1 is differently distributed. The mean time to failure may then be interpreted either as an MTTFF (mean time to first failure), if the mean of X_1 is considered, or as an MTBF if the mean of X_i ($i \geqslant 2$) is considered. The failure rate of an IFR system subject to a general renewal process is shown in Fig. 5.2 (b).

The renewal theory was developed for proper renewals [3] which are equivalent to the replacement of the failed system by a new one. Such renewal processes are typical for non-redundant systems of CFR compounds where each component is replaced on failure.

Taking into account the possible orderings of $R_i(x)$, the positive renewals are defined by the relation :

$$R_1(x) > R_2(x) > \ldots > R_n(x) > \ldots \quad . \qquad (5.3)$$

In the case of CFR and DFR systems, the positive renewals are generated by imperfect maintenance actions which cause positive jumps of the failure rate. It is to be noted that the characteristics of the renewals result from the combined action of the type of system aging and the properties of the maintenance procedures. Thus, in the case of DFR systems, positive renewals arise when the effect of imperfect maintenance is more important than the effect of negative aging. In the case of IFR systems, the renewals may be positive even if the maintenance actions tend to improve the system reliability, if their effect is overwhelmed by the positive aging of the system (Fig. 5.3 (a)).

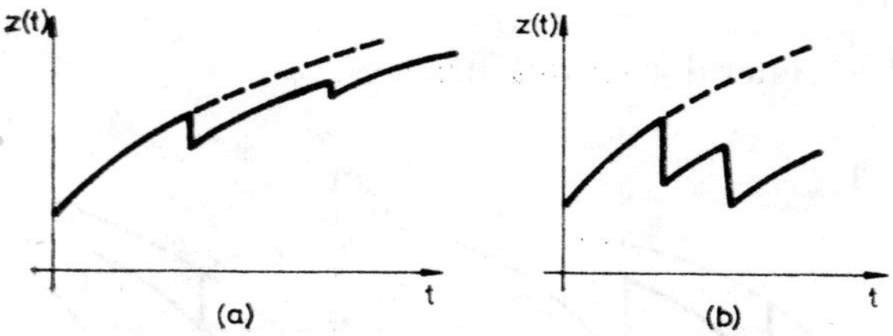

Fig. 5.3. — Effect of renewals on the failure rate of an IFR system :
(a) positive renewals ; (b) negative renewals.

Conversely, the negative renewals are defined by the ordering :

$$R_1(x) < R_2(x) < \ldots < R_n(x) < \ldots \tag{5.4}$$

In the case of CFR and IFR systems, the negative renewals are generated by maintenance procedures which tend to improve the system reliability, causing negative jumps of the failure rate. The renewals are negative for IFR systems if the effect of the maintenance actions is more important than the effect of positive aging. For DFR systems, the renewals may still be negative with imperfect maintenance if the effect of negative aging is more significant.

A limiting case is the "bad as old" renewal, which does not alter the reliability characteristics of the system. Thus, the failure rate maintains its continuous form and the renewals are proper, positive or negative depending on whether the system is CFR, IFR or DFR respectively.

A more realistic approach is to assume that a renewal is more likely to improve the reliability of the system than to degrade it. We define the stochastically negative renewals by the relationship :

$$R_1(x) \overset{st}{<} R_2(x) \overset{st}{<} \ldots \overset{st}{<} R_n(x) \overset{st}{<} \ldots \tag{5.5}$$

which implies that $P(R_{i-1}(x) > y) < P(R_i(x) > y)$ for all x, y and $i = 2, 3, \ldots$

Let us consider a CFR system which is maintained by "bad as old" renewals. The renewal may be considered proper, since the system state after each renewal is the same and the renewal process is simple.

We shall derive the probabilities $P_r(t) = P(N_t = r)$ for all t and r. To do this, we consider the time interval $(0, t + \Delta t)$ and we evaluate

$P_r(t + \Delta t)$. The r renewals occur in $(0, t + \Delta t)$ in the following ways: r renewals in $(0, t)$ and no renewal in $(t, t + \Delta t)$; $r - 1$ renewals in $(0, t)$ and one renewal in $(t, t + \Delta t)$; $r - 2$ renewals in $(0, t)$ and two renewals in $(t, t + \Delta t)$ and so on.

It was shown in Chapter 2 that the failure probability of a CFR system in $(t, t + \Delta t)$ is $\lambda\Delta t + \varepsilon(\Delta t)$ where $\varepsilon(\Delta t)$ tends to zero more rapidly than Δt. On the other hand, the probability of more than one failure in $(t, t + \Delta t)$ is negligible (the property of homogeneity), so the following recursive relation may be readily derived:

$$P_r(t + \Delta t) = P_{r-1}(t)[\lambda\Delta t + \varepsilon(\Delta t)] + P_r(t)[1 - \lambda\Delta t + \varepsilon(\Delta t)]. \quad (5.6)$$

Relation (5.6) may be transformed into the differential equation

$$\frac{dP_r(t)}{dt} = -\lambda P_r(t) + \lambda P_{r-1}(t). \quad (5.7)$$

To solve equation (5.7) we put $P_r(t)$ in the form

$$P_r(t) = v_r(t)e^{-\lambda t}. \quad (5.8)$$

Introducing expression (5.8) into equation (5.7) we get:

$$\frac{dv_r(t)}{dt} = \lambda v_{r-1}(t). \quad (5.9)$$

The initial conditions are

$$v_0(t) = 1$$

$$v_r(0) = 0; \quad r = 2, 3 \ldots \quad (5.10)$$

Solving equation (5.9) with the initial conditions (5.10), we get $v_r(t) = (\lambda t)^r/r!$, so the expression for the renewal process is:

$$P_r(t) = \frac{(\lambda t)^r}{r!}e^{-\lambda t}. \quad (5.11)$$

Expression (5.11) defines the Poisson process. It is often useful to describe a renewal process by the values of the mean and variance of the number of renewals in $(0, t)$.

The mean number of renewals in $(0, t)$ is called the renewal function and, in the case of the Poisson process, has the form

$$H(t) = E_{N_t} = \lambda t. \quad (5.12)$$

According to relation (5.12), the mean number of renewals in any time interval is proportional to the length of the time interval.

Differentiation of the renewal functions (5.12) gives the renewal density:

$$h(t) = H'(t) = \lambda. \quad (5.13)$$

As will be shown below, the renewal density multiplied by Δt is the probability of a renewal occurring in $(t, t + \Delta t)$. In a Poisson renewal process, the renewal density is constant and equal to the failure rate of the CFR system.

The variance of the number of renewals in $(0, t)$ is

$$V(t) = E_{N_t^2} - H^2(t) = (\lambda t)^2. \tag{5.14}$$

In order to predict the evolution of the system, one must know the interval reliability function $R(t, t + x)$ and the mean residual life. Since the CFR system has no memory the probability of success for a period of length x starting at t is:

$$R(t, t + x) = \mathrm{e}^{-\lambda x}. \tag{5.15}$$

The mean residual life, from an arbitrary instant t until the next renewal, is equal to the MTBF:

$$m(t) = \int_0^\infty R(t, t + x)\,\mathrm{d}x = \frac{1}{\lambda}. \tag{5.16}$$

Let us now consider an aging system maintained by "bad as old" renewals. These renewals are positive/negative for IFR/DFR systems, respectively. The renewal process is in this case a non-stationary Poisson process, so that:

$$P(N_t = r) = \frac{\left[\int_0^t z(u)\,\mathrm{d}u\right]^r}{r!}\,\mathrm{e}^{-\int_0^t z(u)\mathrm{d}u}. \tag{5.17}$$

The renewal function and the renewal density are:

$$H(t) = \int_0^t z(u)\,\mathrm{d}u = \ln\frac{1}{R(t)} \tag{5.18}$$

$$h(t) = z(t).$$

The renewal function is equal to the negative logarithm of the reliability function, and the renewal density is equal to the time-dependent failure rate. The mean number of renewals in $(t, t + x)$ may be written:

$$H(t, t + x) = H(t + x) - H(t) = \ln\frac{1}{R(t + x)} - \ln\frac{1}{R(t)} =$$

$$= \ln\frac{R(t)}{R(t + x)} = \ln\frac{1}{R(x/t)}. \tag{5.19}$$

Since the "bad as old" renewals do not alter the reliability characteristics of the system, the interval reliability function and the mean residual life have the same general expressions as those derived in Chapter 2 for non-repairable systems:

$$R(t, t + x) = \frac{R(t + x)}{R(t)};$$

$$m(t) = \frac{1}{R(t)} \int_t^{\infty} R(x)\,du. \tag{5.20}$$

It is important to distinguish, by means of experimental data, whether the renewals are positive, negative or proper. A statistical test was developed in [4] which distinguishes separately between positive and proper renewals, or between negative and proper renewals on the basis of the first three time intervals X_1, X_2, X_3 observed in a sample of n elements. The null hypothesis states that the renewals are proper with the alternative that they are positive:

$$\begin{aligned} H_0 &: R_1(x) = R_2(x) = R_3(x) \\ H_1 &: R_1(x) > R_2(x) > R_3(x). \end{aligned} \tag{5.21}$$

Let x_{i_1}, x_{i_2}, x_{i_3} be the first three time intervals between renewals for the element i of the sample ($i = 1, 2, \ldots, n$). We define the random variable

$$Z_i = \begin{cases} 1 & \text{if } x_{i_1} < x_{i_2} < x_{i_3} \\ 0 & \text{otherwise.} \end{cases} \tag{5.22}$$

When most of the values Z_i are equal to 1, there is strong evidence against the positive character of the renewals and the null hypothesis is accepted. The decision rule is based on the variable $S = \sum_{i=1}^{n} z_i$, and the null hypothesis is rejected if $S \leqslant K$. The limit K is computed from the prescribed value of the risk of the first order α, that is, the probability of rejecting H_0 given that it is true. The probability of rejecting H_0 s given by the binomial cdf:

$$P(S \leqslant K) = \sum_{r=0}^{K} C_n^r p_0^r (1 - p_0)^{n-r}; \tag{5.23}$$

where $p_0 = P(z_i = 1)$.

If H_0 is true, that is the renewals are proper, the probability p_0 is equal to 1/6, since all possible orderings of x_{i_1}, x_{i_2}, x_{i_3} are equally probable and their total number is 6. Expressions (5.23) become in this case :

$$p_0 = P(z_i = 1/H_0) = \frac{1}{6};$$
$$\alpha = \sum_{r=0}^{K} C_n^r (1/6)^r (5/6)^{n-r}. \tag{5.24}$$

The decision limit K results from equation (5.24). The sample n may be computed from the type II risk, but as the relationship between them is difficult to establish, it is reasonable to take as large a sample as possible given the economic constraints.

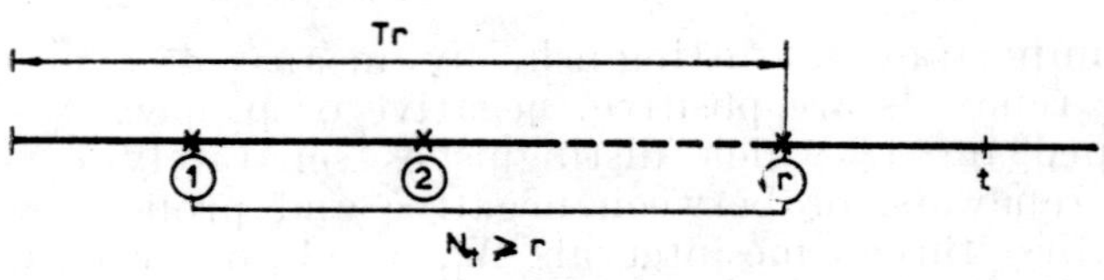

Fig. 5.4. — Derivation of relation (5.25).

The same reasoning holds when we want to distinguish between proper and negative renewals and the decision limit K is the same.

We shall now consider a general system, maintained by proper renewals which bring the system to its initial state. Thus, the renewal process is simple. To derive the expression for $P(N_t = r)$, we note that the number of renewals in $(0, t)$ is greater than or equal to r if and only if the time to the rth renewal is less than t (Fig. 5.4). We may then write :

$$P(N_t \geqslant r) = P(T_r < t). \tag{5.25}$$

Let $K_r(t)$ denote the cdf of T_r and $k_r(t)$ its pdf. For fixed t, the discrete distribution of N_t may be expressed on the basis of the continuous distribution of T_r in the following way :

$$P(N_t = r) = P(N_t \geqslant r) - P(N_t \geqslant r+1) = K_r(t) - K_{r+1}(t) \tag{5.26}$$
$$r = 1, 2, \ldots ; \ K_0(t) = 1.$$

The renewal process is completely specified if the function $K_r(t)$ may be computed from the general reliability measures. This is indeed possible, because all time intervals between renewals are iid according to the pdf $f(x)$ associated with the general reliability model. The time to renewal r is then the sum of the r iid variables $X_1, X_2, \ldots, X_r$ so that the pdf $k_r(t)$ is the r-fold convolution of $f(t)$:

$$k_r(t) = \underbrace{f(t) \otimes f(t) \otimes \ldots \otimes f(t)}_{r}. \tag{5.27}$$

To compute $k_r(t)$, we use the Laplace transform of the expression (5.27) :

$$k_r^*(s) = [f^*(s)]^r. \tag{5.28}$$

Integration of $k_r(t)$ yields the cdf $K_r(t)$, which allows the computation of $P(N_t = r)$ via equation (5.26).

Suppose that the renewals bring the system to a state different from its initial state, the renewal process is general, and a distinction must be made between the distribution $f_1(x)$ of the time to the first failure x_1 and the distribution $f(x)$ of the time intervals between failures, x_i. The relation (5.28) becomes in this case :

$$k^*(s) = f_1^*(s)[f^*(s)]^{r-1}. \tag{5.29}$$

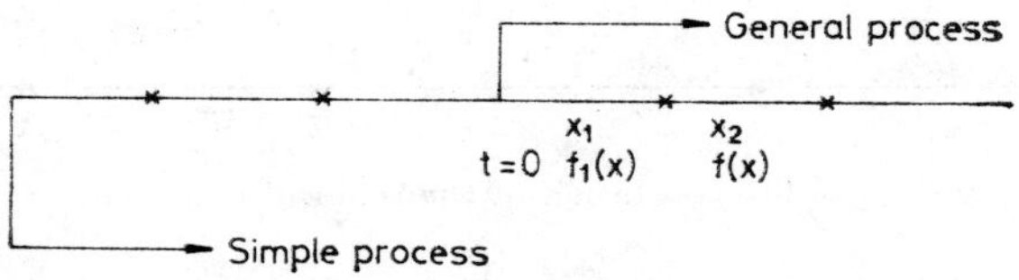

Fig. 5.5. — Interpretation of the general renewal process.

A general renewal process may be obtained when a simple process is observed from an arbitrary instant $t = 0$, as shown in Fig. 5.5.

The fundamental relation (5.26) holds for general renewal processes so the only difference is the computation of $k_r(t)$.

It is often useful to describe the renewal process by synthetical measures like the renewal function and the variance of the number of renewals in $(0, t)$. The renewal function is the mean number of renewals in $(0, t)$, so we may write :

$$H(t) = \sum_{r=0}^{\infty} rP(N_t = r) = \sum_{r=0}^{\infty} r[K_r(t) - K_{r+1}(t)] =$$

$$= K_1(t) + K_2(t) + K_3(t) \ldots = \sum_{r=1}^{\infty} K_r(t). \tag{5.30}$$

The renewal density is the derivative of $H(t)$:

$$h(t) = \sum_{r=1}^{\infty} k_r(t). \tag{5.31}$$

Relation (5.31) shows that $h(t)\Delta t$ may be interpreted as the probability of a renewal occurring in $(t, t + \Delta t)$ irrespective of whether it is the first, the second or the rth, etc., renewal.

In a simple process $k_1(t) = f(t)$ and in a general process $k_1(t) \equiv f_1(t)$.

The Laplace transform of relationship (5.31) gives, in the case of the simple process,

$$h^*(s) = \sum_{r=1}^{\infty} [f^*(s)]^r = \frac{f^*(s)}{1 - f^*(s)} \tag{5.32}$$

and in the case of the general process :

$$h^*(s) = \sum_{r=1}^{\infty} f_1^*(s)[f^*(s)]^{r-1} = \frac{f_1^*(s)}{1 - f^*(s)}. \tag{5.33}$$

Relation (5.33) may be written in the time domain as :

$$h(t) = f_1(t) + h(t) \otimes f(t) = f_1(t) + \int_0^t h(\tau)f(t - \tau)\,\mathrm{d}\tau. \tag{5.34}$$

Fig. 5.6. — Interpretation of the renewal equation.

Equation (5.34) is the renewal equation and may be interpreted in the following way (Fig. 5.6). The probability of a renewal occurring in $(t, t + \Delta t)$ is $h(t)\Delta t$ and is the sum of two terms. The first, $f_1(t)\Delta t$, is the probability that the first renewal occurs in $(t, t + \Delta t)$. The product $h(\tau)f(t - \tau)\mathrm{d}\tau\Delta t$ is the probability that a renewal occurs in $(\tau, \tau + \mathrm{d}\tau)$ and the next renewal occurs in $(t, t + \Delta t)$. Integrating over $\tau \in (0, t)$ we get the second term, which is the total probability that a renewal (not the first one) occurs in $(t, t + \Delta t)$.

Once the renewal density is known, the renewal function is easily computed by integration. In order to evaluate the variance of the number of failures in $(0, t)$, we begin by computing the average of $N_t(N_t + 1)$. By definition, we have :

$$E_{N_t(N_t+1)} = \sum_{r=0}^{\infty} r(r + 1)P(N_t = r) = \sum_{r=1} r(r + 1)[K_r(t) -$$

$$- K_{r+1}(t)] = 2[K_1(t) + 2K_2(t) + 3K_3(t) + 4K_4(t) + \ldots]. \tag{5.35}$$

The Laplace transform of expression (5.35) yields, in the case of the general process,

$$\mathscr{L}E_{N_t(N_t+1)} = \frac{2}{s} f_1^*(s)[1 + 2f^*(s) + 3f^{*2}(s) + 4f^{*3}(s) + \ldots] =$$

$$= \frac{2}{s} \cdot f_1^*(s)\left[\frac{f^*(s)}{1 - f^*(s)}\right]'_{f^*(s)} = \frac{2f_1^*(s)}{s[1 - f^*(s)]^2}. \tag{5.36}$$

In the case of the simple renewal process, relation (5.36) reduces to :

$$\mathscr{L}E_{N_t(N_t+1)} = \frac{2f^*(s)}{s[1 - f^*(s)]^2}. \tag{5.37}$$

The variance of the number of renewals in $(0, t)$ is readily obtained using $E_{N_t(N_t+1)}$ and the renewal function $H(t)$:

$$V_{N_t}(t) = E_{N_t(N_t+1)} - H(t) - H^2(t). \tag{5.38}$$

The renewal function and the variance are useful in the planning of spare parts and, consequently, of support costs.

We shall now consider the example of a system composed of two CFR elements with failure rate λ. One element is active and the other is on stand-by, and is switched when the active element fails. It is assumed that the element on stand-by does not fail, and that the switch is perfect. Renewal is carried out in the event of system failure and restores the system to its initial state.

The time-to-failure of the system is the sum of two iid exponential variables, that is, the times-to-failure of the two elements. Consequently, the pdf of the system time-to-failure is the convolution of two exponential distributions:

$$\begin{aligned} f(t) &= \lambda e^{-\lambda t} \otimes \lambda e^{-\lambda t} = \lambda^2 t e^{-\lambda t} \\ f^*(s) &= \frac{\lambda^2}{(s+\lambda)^2}. \end{aligned} \tag{5.39}$$

The general reliability model of the system is a gamma distribution with parameters $(2, \lambda)$. The mean time between renewals of the system is then $2/\lambda$.

Since the process is simple, the renewal density is

$$\begin{aligned} h^*(s) &= \frac{f^*(s)}{1 - f^*(s)} = \frac{\lambda^2}{s(s+2\lambda)}; \\ h(t) &= \frac{\lambda}{2}(1 - e^{-2\lambda t}). \end{aligned} \tag{5.40}$$

The renewal function is:

$$\begin{aligned} H(t) &= \frac{\lambda t}{2} - \frac{1}{4}(1 - e^{-2\lambda t}) \\ H(t, t+x) &= \frac{\lambda x}{2} - \frac{1}{4}[e^{-2\lambda t} - e^{-2\lambda(t+x)}]. \end{aligned} \tag{5.41}$$

It may be noted that the renewal density tends asymptotically to the inverse of the mean time between renewals. The renewal function tends asymptotically to the line $y = \frac{\lambda t}{2} - \frac{1}{4}$, the slope of which is equal to the inverse of the mean time between renewals. This means that, after a sufficient time has passed, the probability of a renewal around t is constant and the average number of renewals in $(t, t + x)$ is proportional to the length x of the time interval (Fig. 5.7).

The asymptotic behaviour outlined above is quite general. To prove this, we shall take the limit of the general relation (5.33) for $t \to \infty$:

$$\lim_{t\to\infty} h(t) = \lim_{s\to 0} sh^*(s) = \lim_{s\to\infty} \frac{sf_1^*(s)}{1 - f^*(s)} = \lim_{s\to 0} \frac{f_1^*(s)}{R^*(s)} = \frac{1}{m}. \tag{5.42}$$

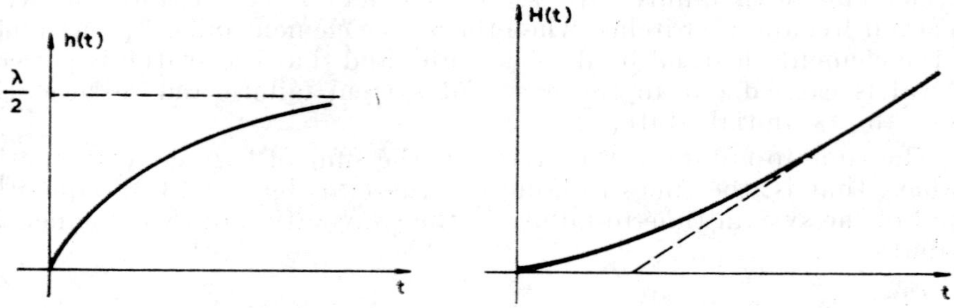

Fig. 5.7. — Renewal function and renewal density of a stand-by system.

A renewal process with a constant renewal density is stationary. From relation (5.42) it follows that any renewal process is asymptotically stationary. The renewal function in $(t, t + x)$ tends to :

$$\lim_{t\to\infty} H(t, t + x) = \lim_{t\to\infty}[H(t + x) - H(t)] = \frac{x}{m}\cdot \tag{5.43}$$

This result is known as the Blackwell-Smith renewal theorem [2]. The renewal process of CFR systems is stationary from the beginning.

The stationary behaviour of the renewal processes justifies the usual way of computing the average number of renewals in a time interval by dividing the length of the time interval by the MTBF. Such a procedure is correct only in the stationary state of the renewal process, and gives erroneous results when used for assessing the maintenance resources necessary during the transitory state.

In most cases, when the asymptotic formula (5.43) is used in the transitory state, the maintenance resources are over-estimated. It is then more economical to perform an accurate evaluation of the renewal function, especially in the case of high reliability systems when the transitory time is large.

Prediction of the system behaviour should be possible at each instant. At the instants of failure, the reliability function $R(t)$ and the mean time between failures are sufficient to predict the instant when the next system failure is bound to occur. In order to make such a prediction at an arbitrary instant t, we shall consider the time between failures divided by the prediction instant into two subintervals $U(t)$ and $V(t)$, which represent the actual age of the system and its residual life, respectively (Fig. 5.8). In order to predict at t the next system failure, we must derive the distribution of the residual life $V(t)$.

If t precedes the first system failure, the pdf of $V(t)$ is $f_1(t + v)$. If t is situated between two successive failures, we assume that the most recent renewal has occurred around $t - u$ and that the next renewal is bound to occur around $t + v$. The probability of the latest renewal occurring around $t - u$ is $h(t - u)\,du$. As the system is new at $t - u$,

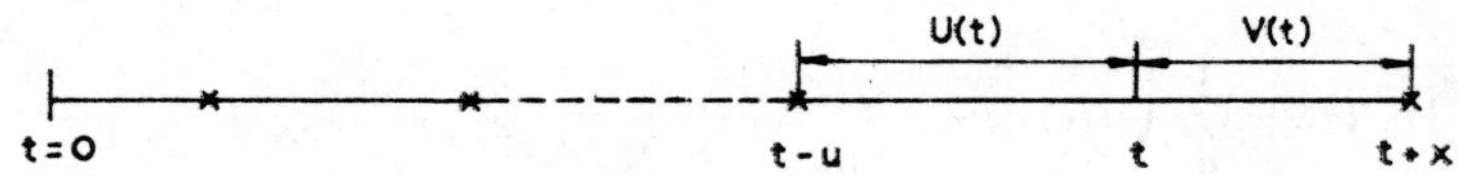

Fig. 5.8. — System age and residual life.

the conditional pdf that the next renewal occurs around $t + v$ is $f(u + v)$. According to the total probability formula, summing for all $u \in (0, t)$ we get the pdf of $V(t)$:

$$f_v(v, t) = f_1(t + v) + \int_0^t h(t - u) f(u + v)\,du. \qquad (5.44)$$

The interval reliability function may be readily derived from the pdf of the residual life:

$$R(t, t + x) = P(V(t) \geqslant x) = \int_x^\infty f_v(v, t)\,dv =$$

$$= R_1(t + x) + \int_0^t h(t - u) R(u + x)\,du. \qquad (5.45)$$

Thus we can state at time t that the next failure will not occur until $t + x$, with a probability $R(t, t + x)$ given by relation (5.45).

The mean residual life $m(t)$ may also be derived, by introducing expression (5.45) in :

$$m(t) = \int_0^\infty R(t,\ t + x)\,dx. \qquad (5.46)$$

Comparing the interval reliability (5.45) with the interval reliability of a non-repairable system $R(x/t)$, we notice that the interval reliabilities differ even if the reliability function $R(x)$ is the same, because of the influence of the maintenance procedures.

It should also be noticed that the interval reliability of a non-repairable system is conditioned by the system survival up to t, while the interval reliability of a repairable system is a total (unconditioned) probability. For $t = 0$, $R(t, t + x)$ reduces to $R_1(x)$ and for $x = 0$ it is equal to 1, that is the probability of the system being operational at each instant is unity as the renewal durations were neglected.

Prediction of the next failure is much easier in the stationary state. For $t \to \infty$, the distribution of the residual life becomes :

$$\lim_{t\to\infty} f_v(v, t) = \frac{1}{m}\int_x^\infty f(t)\,\mathrm{d}t = \frac{R(v)}{m}. \tag{5.47}$$

The interval reliability and the mean residual life are, respectively,

$$\begin{aligned} &\lim_{t\to\infty} R(t, t + x) = R_s(x) = \frac{1}{m}\int_x^\infty R(t)\,\mathrm{d}t\,; \\ &\lim_{t\to\infty} m(t) = E_v = \frac{1}{m}\int_0^\infty xR(x)\,\mathrm{d}x = \frac{1}{2m}\int_0^\infty x^2 f(x)\,\mathrm{d}x. \end{aligned} \tag{5.48}$$

In the particular case of a CFR system, the renewal process is simple and stationary from the beginning, so that

$$h(t) = \lambda\,; \quad f_1(x) = f(x) = \lambda \mathrm{e}^{-\lambda x}. \tag{5.49}$$

Taking into account relation (5.49), the expressions (5.44) to (5.48) become :

$$\begin{aligned} &f_v(v, t) = \lambda \mathrm{e}^{-\lambda v} = \lim_{t\to\infty} f_v(v, t) \\ &R(t, t + x) = \mathrm{e}^{-\lambda x} = \lim_{t\to\infty} R(t, t + x) \\ &m(t) = \frac{1}{\lambda} = \lim_{t\to\infty} m(t). \end{aligned} \tag{5.50}$$

On the other hand, the time interval between failures $x(t) = U(t) + V(t)$ is exponentially distributed, with mean value $1/\lambda$. This means that, in an arbitrary section of an exponential variable, we find another exponential variable with the same mean. The part seems equal to the whole. In order to clarify this apparent paradox, we shall consider also the variable $U(t)$ which is the actual age of the system at time t.

For $u < t$, the probability density function of $U(t)$ is the product of the renewal density at $t - u$ and the reliability function in $(t - u, t)$, which is $R(u)$.

The system age $U(t)$ is t only if it has not yet failed up to t, an event with probability $R_1(t)$. The distribution $f_U(u, t)$ of $U(t)$ is then :

$$f_U(u, t) = \begin{cases} h(t-u)R(u) & u < t \\ R_1(t) & u = t \\ 0 & u > t. \end{cases} \tag{5.51}$$

In the stationary state, we get :

$$\lim_{t \to \infty} f_U(u, t) = \frac{R(u)}{m}. \tag{5.52}$$

The asymptotic mean of $U(t)$ is then :

$$E_U = \frac{1}{m}\int_0^\infty uR(u)\mathrm{d}u = \frac{1}{2m}\int_0^\infty x^2 f(x)\,\mathrm{d}x = E_V. \tag{5.53}$$

The paradox may now be considered from a more general point of view. On the one hand, the time between failures $X(t)$ is distributed according to $f(x)$ with mean $m = \int_0^\infty xf(x)\,\mathrm{d}x$. On the other hand, it is equal to the sum of $U(t)$ and $V(t)$, so that its mean should be $\frac{1}{m}\int_0^\infty x^2 f(x)\,\mathrm{d}x$. The results are clearly contradictory.

The solution of the paradox stems from the observation that the distribution of the time interval $X(t)$ is conditioned by the system being new at the beginning of the interval. We shall derive the unconditional distribution of $X(t)$.

The probability of a system renewal around τ is $h(\tau)\mathrm{d}\tau$. As the system is new at τ, the next failure occurs around $\tau + x$ with probability $f(x)\mathrm{d}x$. If $x < t$, the instant t is situated between two successive failures, so that τ is at most t and at least $t - x$. According to the total probability formula, we get the unconditional distribution $g(x, t)$ of $X(t)$:

$$g(x, t) = \int_{t-x}^{t} h(\tau)f(x)\,\mathrm{d}\tau = f(x)[H(t) - H(t-x)]; \quad x < t. \tag{5.54}$$

If $x > t$, it is possible that t should precede the first system failure. In this case the distribution of $X(t)$ is $f_1(x)$. If t is situated between τ and $\tau + x$, the distribution of $X(t)$ is $f(x)$ and it is conditioned by a renewal occurring around τ, with probability $h(\tau)\,\mathrm{d}\tau$. By the total probability formula, taking into account that $\tau \in (0, t)$, the unconditional distribution $g(x, t)$

of $X(t)$ for $x > t$ is :

$$g(x, t) = f_1(x) + \int_0^t h(\tau)f(x)\,d\tau = f_1(x) + f(x)H(t); \quad x > t. \tag{5.55}$$

In the stationary state, we get the asymptotic expression :

$$\lim_{t\to\infty} g(x, t) = \frac{xf(x)}{m}. \tag{5.56}$$

The unconditional mean time between failures is then :

$$E_X = \frac{1}{m}\int_0^\infty x^2 f(x) = E_U + E_V. \tag{5.57}$$

The solution of the paradox shows that the reliability model is dependent on the observer [5]. If the observer is situated at a renewal instant, the distribution of the time between failures is $f(x)$, while if the observer is situated at an arbitrary instant between failures, the distribution of the time between failures is $g(x, t)$.

In the case of a CFR system, the distribution of the system age at t is :

$$f_U(u, t) = \begin{cases} \lambda e^{-\lambda u} & u < t \\ e^{-\lambda t} & u = t \\ 0 & u > t. \end{cases} \tag{5.58}$$

In the stationary state, the distribution (5.58) becomes :

$$\lim_{t\to\infty} f_U(u, t) = \lambda e^{-\lambda u}. \tag{5.59}$$

It follows from relations (5.50) and (5.59) that an arbitrary section of an exponential variable with mean $1/\lambda$ gives two exponential variables, each having the mean value $1/\lambda$. This is explained by the fact that the section of the time interval changes its distribution from $\lambda e^{-\lambda}$ to :

$$g(x, t) = \begin{cases} \lambda^2 x\, e^{-\lambda x}; & x < t \\ \lambda e^{-\lambda x}(1 + \lambda t); & x > t. \end{cases} \tag{5.60}$$

In the stationary state, $g(x, t) = \lambda^2 x e^{-\lambda x}$ and $E_x = 2/\lambda = E_U + E_V$. In the transitory state, we have :

$$E_U = \int_0^t \lambda u e^{-\lambda u}\,du + te^{-\lambda t} = \frac{1}{\lambda}(1 - e^{-\lambda t}); \quad E_V = \frac{1}{\lambda}. \tag{5.61}$$

Taking the mean of $X(t)$ with respect to $g(x, t)$ we get :

$$E_X = \int_0^\infty x g(x, t)\, dx = \frac{2}{\lambda} - \frac{1}{\lambda} e^{-\lambda t} = E_U + E_V. \tag{5.62}$$

The apparent paradox vanishes as soon as the relativistic feature of the reliability model is taken into account, which means that the time-to-failure distribution is modified by the presence of the observer at an arbitrary instant t between two successive failures. In the example of the CFR system, the time between failures is exponentially distributed with mean $1/\lambda$ for an observer present at the instant of the last failure, but is distributed according to the law (5.60) with mean (5.62) for an observer present at an arbitrary instant t between failures.

The residual life allows an alternate definition of the stationary renewal process. We have seen that a general renewal process may be obtained from an arbitrary section of a simple process (Fig. 5.5). It follows that the time to the first failure X_1 in the general process is precisely the residual life $V(t)$ in the simple process. As the distribution of the residual life may be computed from the general reliability model by relation (5.44), the characteristics of the general process may be readily determined. Moreover, if the simple process is sectioned long after it has begun, the general process is stationary and the distribution of X_1 has the asymptotic expression (5.47). The stationary process may then be defined as a general process, with the time to the first failure distributed according to $R(x)/m$. It may easily be shown, using expression (5.33), that when $f_1(x) = R(x)/m$ the renewal density is constant and equal to $1/m$.

The renewal process describes, from a global point of view, the reliability of systems with "good as new" repairs. The renewal function allows an evaluation of support costs, while the interval reliability and the mean residual life are useful for predicting the system behaviour.

We shall now consider other types of renewals : positive, negative, and stochastically positive or negative. If the renewals do not have any constant feature, that is, positive and negative renewals occur at random, the system reliability may only be described separately, for each time interval between renewals. It is not possible to predict the system behaviour beyond the next renewal.

If the renewals maintain a constant pattern for a sufficiently long time span (the renewals are all positive or all negative), a global model of the renewal process may be established. To this end, it is necessary to find the relationship between the reliability functions for different time intervals between renewals. Such a relationship is derived in [6], where the effect of each positive (or negative) renewal is described by a constant ratio of the areas of the failure rate on two successive time intervals between renewals. The ratio is greater than unity for positive renewals and less than unity for negative renewals.

Another way to describe the renewal effect is to adopt a certain distribution family for all time intervals between renewals such that one or more of its parameters are specified functions of the interval number. For instance, a CFR system with negative renewals has all the time intervals between renewals exponentially distributed with parameters $\lambda(i)$, where $\lambda(i)$ is a decreasing function of the interval number i. This is precisely the case of software systems, where the failure rate is constant for each time interval between renewals and is proportional to the number of latent errors in the program. If an error is removed at each failure, the failure rate decreases by the same factor, so that it is a decreasing function of the number of failures.

The effect of renewals may be better described by the concept of stochastically negative (or positive) renewals.

Loosely speaking, a renewal is stochastically negative if it is more likely to improve the system reliability than to degrade it. It is then possible for the system reliability to remain unchanged or become worse after a renewal, but most of the renewals tend to improve it. The converse is true for stochastically positive renewals.

The notion of stochastically negative (or positive) renewal may be properly defined by associating probability density functions with different time intervals between renewals. The cumulative density function associated with the reliability during time interval i is decreasing with i for stochastically negative renewals and increasing with i for stochastically positive renewals. We then have

$$\begin{aligned} &P(R_{i+1} < r) < P(R_i < r) \\ &\forall\, r \geqslant 0\,;\; i = 1, 2, \ldots \end{aligned} \tag{5.63}$$

for stochastically negative renewals and

$$\begin{aligned} &P(R_{i+1} < r) > P(R_i < r) \\ &\forall\, r > 0\,;\; i = 1, 2, \ldots \end{aligned} \tag{5.64}$$

for stochastically positive renewals.

5.2 Negative renewals

We shall consider a CFR system subject to negative renewals. The failure rate of the system is constant between two successive renewals, and is a decreasing function of the number of renewals (Fig. 5.9).

The time intervals between renewals are independent, and distributed according to exponential laws with different parameters λ_k. Let us adopt for $\lambda(k)$ the simple expression

$$\lambda(k) = (M - k + 1)\varphi \tag{5.65}$$

where M and φ are constants.

The reliability function for time interval k is then:

$$R_k(x) = \exp\left[-(M - k + 1)\varphi x\right] \tag{5.66}$$

and the mean time between renewals $(k - 1)$ and k is:

$$m_k = \frac{1}{(M - k + 1)\varphi}. \tag{5.67}$$

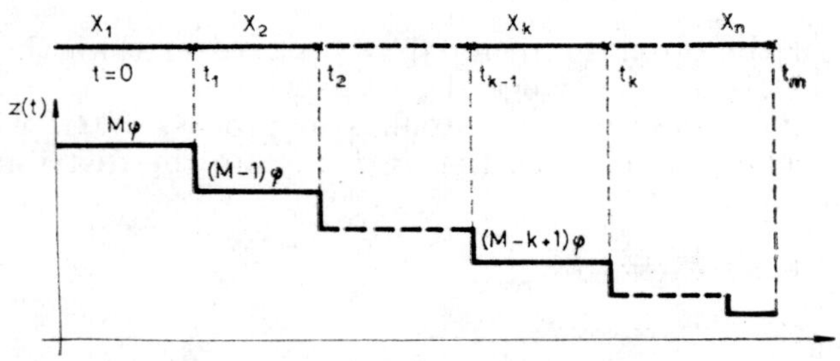

Fig. 5.9. — CFR system with negative renewals

Expressions (5.66) and (5.67) are sufficient to predict the system behaviour when the observer is situated at a failure time, say t_{k-1}. When the observer is situated at a time instant $t \in (t_{k-1}, t_k)$, prediction of the next failure is based on the interval reliability function $R(t, t + x)$ and the mean residual life $m(t)$. As the system has no memory in the time interval between failures, we may write:

$$R(t, t + x) = \exp[-(M - k + 1)\varphi x];$$

$$m(t) = \frac{1}{(M - k + 1)}; \tag{5.68}$$

$$t_{k-1} \leqslant t < t + x \leqslant t_k; \quad k = 1, 2, \ldots$$

The mean residual life is then equal to the mean time between failures $k - 1$ and k. This is the same apparent paradox encountered in the previous Section, that is, a part is equal to the whole, on average. The paradox is easily solved if we notice that the distribution of x_k is $f_k(x_k) = \lambda_k \exp[-\lambda_k x_k]$ only for an observer situated at time instant t_{k-1}. The observer situated at t knows that the system has not failed in time interval (t_{k-1}, t) so that, for him, the distribution $g(t, x_k)$ of X_k is conditional on the system survival in (t_{k-1}, t):

$$g_k(t, x_k) = \frac{f_k(x_k)}{R_k(t - t_{k-1})} =$$

$$= \begin{cases} \lambda_k \exp[-\lambda_k(x_k - t + t_{k-1})]; & x_k > t - t_{k-1} \\ 0 & ; \quad x_k < t - t_{k-1}. \end{cases} \tag{5.69}$$

Taking the mean of x_k, we have now:

$$Ex_k = \int_0^\infty x_k g_k(t, x_k)\mathrm{d}x_k = t - t_{k-1} + \frac{1}{\lambda_k} =$$

$$= t - t_{k-1} + \frac{1}{(M - k + 1)\varphi}. \tag{5.70}$$

The mean x_k for the observer situated at t is greater than the mean residual life. The part is no longer equal to the whole.

We shall now describe the stochastic process $N(t)$, where $N(t)$ is the number of renewals in $(0, t)$. Let $P_r(t)$ denote the distribution of $N(t)$ for fixed t, so that:

$$P(N(t) = r) = P_r(t). \tag{5.71}$$

It is clear that:

$$\begin{aligned} &P_0(t) = \mathrm{e}^{-\lambda_1 t} = \mathrm{e}^{-M\varphi t} \\ &P_r(0) = 0\,; \quad r = 1, 2, \ldots \end{aligned} \tag{5.72}$$

The probability of renewal k occurring in time interval $(t, t + \Delta t)$ is $\lambda_k \Delta t$, where $\lambda_k = (M - k + 1)\varphi$. Considering the time interval $(0, t + \Delta t)$ the renewal process is:

$$\begin{aligned} &P_r(t + \Delta t) = P_r(t)(1 - \lambda_{r+1}\Delta t) + P_{r-1}(t)\lambda_r\Delta t \\ &r = 1, 2, \ldots \end{aligned} \tag{5.73}$$

Equation (5.73) is easily transformed into the differential equation:

$$\frac{\mathrm{d}P_r(t)}{\mathrm{d}t} = -\lambda_{r+1}P_r(t) + \lambda_r P_{r-1}(t)\,; \quad r = 1, 2, \ldots \tag{5.74}$$

where: $\lambda_r = (M - r + 1)\varphi$; $\lambda_{r+1} = (M - r)\varphi$.

Solving the differential equations (5.74) with initial conditions (5.71,) we get [7]:

$$P_r(t) = C_M^r(\mathrm{e}^{-\varphi t})^{M-r}(1 - \mathrm{e}^{-\varphi t})^r\,; \quad r = 0, 1, 2 \ldots \tag{5.75}$$

The number of renewals in $(0, t)$ is binomially distributed with parameters $1 - \mathrm{e}^{-\varphi t}$ and M. The renewal function $H(t)$ is then the mean of $N(t)$, that is:

$$H(t) = M(1 - \mathrm{e}^{-\varphi t}) \tag{5.76}$$

and the renewal density is:

$$h(t) = H'(t) = M\varphi\mathrm{e}^{-\varphi t} \tag{5.77}$$

The variance of the number of renewals is:

$$V(t) = M\mathrm{e}^{-\varphi t}(1 - \mathrm{e}^{-\varphi t}). \tag{5.78}$$

The renewal process with negative renewals is completely determined, as it provides predictions for the system evolution at any time instant and the average number of renewals for any time interval is known.

Expression (5.65) adopted for the failure rate implies that the maximum number of system renewals is M. This restriction may be easily interpreted for the case of software systems. A software system is a CFR system, as it does not wear out. Let M be the total number of latent errors in the system. A failure occurs when a specific set of input data causes an error to become apparent. It is assumed that an error is removed at each renewal, so that the number of renewals cannot exceed M. The failure rate is proportional to the number of remaining errors, with factor φ. In this application, it is interesting to find the distribution of the remaining errors $Y(t) = M - N(t)$. From expression (5.75) it is easy to derive :

$$Q_y(t) = P[Y(t) = y] = C_M^{M-y}(\mathrm{e}^{-\varphi t})(1 - \mathrm{e}^{-\varphi t})^{M-y}. \tag{5.79}$$

Thus, the number of remaining errors is binomially distributed with parameters M and $\mathrm{e}^{-\varphi t}$. The mean number of remaining errors at time t is :

$$E_{Y(t)} = Me^{-\varphi t}. \tag{5.80}$$

The probability that all errors are removed in $(0, t)$ is :

$$P(Y(t) = 0) = (1 - \mathrm{e}^{-\varphi t})^M. \tag{5.81}$$

From equation (5.81) we may derive the time to test a software system so that, with probability P, no errors are left :

$$t = \frac{1}{\varphi}\ln\frac{1}{1 - P^{1/M}}. \tag{5.82}$$

Practical applications of the negative renewal process rest on the possibility of estimating parameters M and φ from experimental data. Let $x_1, x_2, \ldots, x_n$ be observed time intervals between renewals. The distribution of X_k is :

$$f(x_{k/M,\varphi}) = (M - k + 1)\varphi \exp[-(M - k + 1)\varphi x_k];$$

$$k = 1, 2, \ldots, n. \tag{5.83}$$

As $X_1, X_2, \ldots, X_n$ are independent random variables, their joint distribution is :

$$f(x_1, x_2, \ldots, x_n/M, \varphi) =$$

$$= \prod_{k=1}^{n} \varphi(M - k + 1)\exp[-(M - k + 1)\varphi x_k]. \tag{5.84}$$

Taking the logarithm of expression (5.84), we get the likelihood function :

$$L(x_1, x_2, \ldots, x_n/M, \varphi) = \sum_{k=1}^{n} \ln(M - k + 1) -$$

$$- \varphi \sum_{k=1}^{n} (M - k + 1)x_k + n \ln \varphi. \tag{5.85}$$

The estimates of M and φ are the values that maximize the likelihood function, so that :

$$\frac{\partial L(x_1, x_2, \ldots, x_n/M, \varphi)}{\partial M} = 0; \quad \frac{\partial L(x_1, x_2, \ldots, x_n/M, \varphi)}{\partial \varphi} = 0. \tag{5.86}$$

The system (5.86) yields :

$$\hat{\varphi} = \frac{n}{Mt_n - \sum_{k=1}^{n} (k - 1)x_k};$$

$$\sum_{k=1}^{n} \frac{1}{M - k + 1} = \frac{nt_n}{Mt_n - \sum_{k=1}^{n} (k - 1)x_k} \tag{5.87}$$

where $t_n = \sum_{k=1}^{n} x_k$.

The model described may easily be adapted for the case of positive renewals if the failure rate λ_k is an increasing function of the interval number.

In the case of software systems, this would imply that each intervention feeds into the program more errors than it corrects.

5.3 Stochastically negative renewals

The stochastically negative renewal process is by far the most interesting, as it is more likely that such a renewal will upgrade the system reliability than degrade it. Moreover, any system reliability analysis is primarily concerned with systems in their early life cycle, when their aging is negative, so that even minimal (bad as old) repairs may be interpreted as stochastically negative renewals.

The time-to-failure distribution may be approximated by a piecewise exponential function. The failure rate is constant, with jumps at the instants of failure. In order to approximate better the true time-to-failure distribution, supplementary discontinuities of the failure rate are inserted between two successive renewals. Such discontinuities are called pseudo-failures. Specification of the parameters of the piecewise exponential model must take into account the type of aging and the type of renewals.

Let $f_{a_i}(\lambda_i)$ be the a-priori distribution associated with the parameter λ_i. According to the definition (5.63) of stochastically negative renewals, we get :

$$F_{a_{i+1}}(\lambda) > F_{a_i}(\lambda) \qquad \forall\, \lambda > 0, \quad i = 1, 2, \ldots \tag{5.88}$$

The system is assumed to be of the DFRA type, so that relation (5.88) is true for any i, irrespective of the fact that the discontinuity of the failure rate is generated by a failure or by a pseudo-failure.

According to the methods of bayesian inference, an a-priori distribution $f_{a_i}(\lambda)$ of the gamma type is adopted :

$$f_{a_i}(\lambda) = \frac{\beta_i(\beta_i\lambda)^{\alpha-1}e^{-\beta_i\lambda}}{\Gamma(\alpha)}. \tag{5.89}$$

The parameter α is the same for any exponentially distributed time interval, while β_i is an increasing function of the interval number i, such that condition (5.88) is satisfied. Indeed, the cdf associated with λ_i is :

$$F_{a_i}(\lambda) = \int_0^{\lambda} f_{a_i}(\lambda)\mathrm{d}\lambda = \frac{1}{\Gamma(\alpha)}\int_0^{\beta_i\lambda} u^{\alpha-1}e^{-u}\mathrm{d}u. \tag{5.90}$$

It is obvious that $F_{a_i}(\lambda)$ is increasing if β_i is increasing. It is assumed that a-priori information completely specifies the function $\beta(i)$. For many systems we may adopt the form :

$$\beta(i) = \ln(i+1). \tag{5.91}$$

Nothing is specified a-priori as far as parameter α is concerned so that α must be estimated from experimental data alone. In the bayesian framework, this implies associating with α a uniform a-priori distribution :

$$\psi_a(\alpha) = \begin{cases} 1 & \alpha \geqslant 0 \\ 0 & \alpha < 0. \end{cases} \tag{5.92}$$

First, a Bayes inference concerning α will be obtained. Secondly, we shall obtain the a-posteriori distributions associated with λ_i, so that predictions on future system behaviour can be provided [9].

The available data are the time intervals between failures $x_1, x_2, \ldots, x_n$. We assume first that no pseudo-failures are inserted, that is, the only discontinuities of the failure rate are the failure instants $t_1, t_2, \ldots, t_n$. The prediction is made by an observer situated at time instant t_n.

According to the initial assumptions, the time intervals $\{X_i ; i = 1, 2, \ldots, n\}$ are exponentially distributed, that is :

$$f_i(x_i/\lambda_i) = \lambda_i e^{-\lambda_i x_i}. \tag{5.93}$$

Taking expression (5.89) into account we find, from the total probability formula:

$$f_i(x_i/\alpha) = \int_0^\infty f_i(x_i/\lambda_i) f_{a_i}(\lambda_i) \mathrm{d}\lambda_i = \frac{\alpha \theta_i^\alpha}{\beta_i + x_i}; \tag{5.94}$$

where $\theta_i = 1 + \frac{x_i}{\beta_i}$.

As $x_1, x_2, \ldots, x_n$ are independent random variables, their joint probability density function is:

$$L(x_1, x_2, \ldots, x_n/\alpha) = \sum_{i=1}^{n} f_i(x_i/\alpha) = \frac{\alpha^n \theta^{-\alpha}}{\prod_{i=1}^{n} (\beta_i + x_i)} \tag{5.95}$$

where $\theta = \prod_{i=1}^{n} \theta_i$.

We introduce relations (5.92) and (5.95) into the Bayes equation in order to obtain the a-posteriori distribution of α:

$$\psi_p(\alpha) = \frac{\psi_a(\alpha) L(x_1, x_2, \ldots, x_n/\alpha)}{\int_0^\infty \psi_a(\alpha) L(x_1, x_2, \ldots, x_n/\alpha) \mathrm{d}\alpha} =$$

$$= \frac{\alpha^n \theta^{-\alpha} (\ln \theta)^{n+1}}{\Gamma(n+1)} = \frac{\eta (\eta\alpha)^n \mathrm{e}^{-\eta\alpha}}{\Gamma(n+1)}, \tag{5.96}$$

where $\eta = \ln \theta = \sum_{i=1}^{n} \ln\left(1 + \frac{x_i}{\beta_i}\right)$.

The a-posteriori distribution of α is of the gamma type, with parameters $n+1$ and η. The Bayes estimate of α is the posterior mean:

$$\hat{\alpha} = \frac{n+1}{\eta}. \tag{5.97}$$

The Bayes estimate of λ_{n+1} is the posterior mean of the gamma distribution (5.89), that is:

$$\hat{\lambda}_{n+1} = \frac{\hat{\alpha}}{\beta_{n+1}} = \frac{n+1}{\eta \beta_{n+1}}. \tag{5.98}$$

In general, the Bayes estimate at time t_n of the failure rate for any future time interval $n+k$ is:

$$\hat{\lambda}_{n+k} = \frac{\hat{\alpha}}{\beta_{n+k}} = \frac{n+1}{\eta \beta_{n+k}}. \tag{5.99}$$

If one wants to predict at t_n the testing time necessary to attain a specified low failure rate λ_0, expression (5.99) is set equal to λ_0 and the number $k-1$ of future required renewals is obtained. This is a useful way to determine the burn-in duration but has the disadvantage that the result is expressed as the number of renewals rather than in time units.

The mean time between failures n and $n+1$ is then :

$$\hat{m}_{n+1} = \frac{\eta\beta_{n+1}}{n+1} \tag{5.100}$$

and the mean time between failures $n+k-1$ and $n+k$, estimated at t_n, is :

$$\hat{m}_{n+k} = \frac{\eta\beta_{n+k}}{n+1}. \tag{5.101}$$

To complete the prediction, we have to estimate the reliability functions $R_{n+1}(x)$ and $R_{n+k}(x)$. As x_{n+1} and x_{n+k} are exponentially distributed, we have :

$$\begin{aligned} R_{n+1}(x) &= \exp[-\lambda_{n+1}x]; \\ R_{n+k}(x) &= \exp[-\lambda_{n+k}x]. \end{aligned} \tag{5.102}$$

The Bayes estimates of the reliability functions (5.102) are obtained by taking their means over the posterior distributions of λ_{n+1} and λ_{n+k} respectively.

$$\begin{aligned} R_{n+1}(x) &= \int_0^\infty e^{-\lambda x}\mathrm{d}\lambda \int_0^\infty f_{a_{n+1}}(\lambda/\alpha)\psi_p(\alpha)\mathrm{d}\alpha = \\ &= \int_0^\infty \psi_p(\alpha)\mathrm{d}\alpha \int_0^\infty e^{-\lambda x} f_{a_{n+1}}(\lambda/\alpha)\mathrm{d}\lambda = \left[\frac{\eta}{\eta + \ln\left(1+\frac{x}{\beta_{n+1}}\right)}\right]^{n+1}; \\ R_{n+k}(x) &= \int_0^\infty e^{-\lambda x}\mathrm{d}\lambda \int_0^\infty f_{a_{n+k}}(\lambda/\alpha)\psi_p(\alpha)\mathrm{d}\alpha = \\ &= \int_0^\infty \psi_p(\alpha) \int_0^\infty e^{-\lambda x} f_{a_{n+k}}(\lambda/\alpha)\mathrm{d}\lambda = \left[\frac{\eta}{\eta + \ln\left(1+\frac{x}{\beta_{n+k}}\right)}\right]^{n+1}. \end{aligned} \tag{5.103}$$

Expressions (5.103) are used to predict at time t_n the reliability functions after n and $n+k-1$ failures, respectively. A useful prediction formula is given by the F-percentile of the time interval from failure n to failure

$n + 1$. Setting $R_{n+1} = 1 - F$, we get :

$$x_{n+1;F} = \beta_{n+1}[\theta^{(1-F)^{-\frac{1}{n+1}}-1} - 1]. \tag{5.104}$$

At each failure instant, the time to the next failure may then be computed for a specified probability of failure, F.

The accuracy of the predictions made may be improved by inserting one or more discontinuities of the failure rate between two failures [10]. In order to do this, the median of x_{n+1} is computed, setting $F = 0.5$ in expression (5.104) :

$$x_{n+1;0.5} = \beta_{n+1}[\theta^{2^{\frac{1}{n+1}}-1} - 1]. \tag{5.105}$$

When system failure $(x + 1)$ occurs, the actual time interval x_{n+1} is compared to the prediction of its median (5.105). There are two cases :

a) $x_{n+1} \leqslant x_{n+1;0.5}$. No additional discontinuity is inserted in the time interval $n + 1$. The value x_{n+1} is used for the next prediction.

b) $x_{n+1} > x_{n+1;0.5}$. A discontinuity is inserted in the time interval $n + 1$, according to Fig. 5.10. The additional discontinuity point of the failure rate is called a pseudo-failure and it is treated partly like the other failures in the predictions. The pseudo-failure inserted in case (b) divides the time interval x_{n+1} into two unequal parts $x_{n+1}^{(1)} = x_{n+1;0.5}$ and $x_{n+1}^{(2)} = x_{n+1} - x_{n+1}^{(1)}$, the value β_{n+1} is associated with the first subinterval, and the value β_{n+2} with the second. The first subinterval is used together with the first n time intervals to predict the median of $x_{n+1}^{(2)}$:

$$\begin{aligned} x_{n+1;0.5}^{(2)} &= \beta_{n+2}(\theta^{2^{\frac{1}{n+1}}-1} - 1); \\ \theta &= \left(1 + \frac{x_{n+1}^{(1)}}{\beta_{n+1}}\right) \prod_{i=1}^{n} \left(1 + \frac{x_i}{\beta_i}\right). \end{aligned} \tag{5.106}$$

If the predicted value $x_{n+1;0.5}^{(2)}$ is less than the actual value $x_{n+1}^{(2)}$, another pseudo-failure is inserted in $x_{n+1}^{(2)}$. Otherwise, the time interval x_{n+2} is predicted, using the failures and also the pseudo-failures.

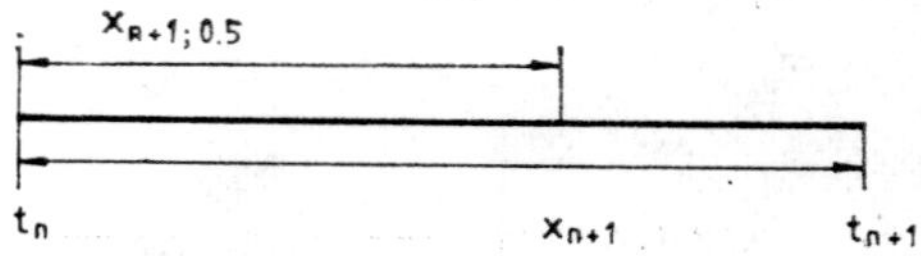

Fig. 5.10. — The insertion of a pseudo-failure.

Suppose that, up to failure n, there were j pseudo-failures inserted in the first n time intervals. Let $\{x_i ;\ i = 1, 2, \ldots, n + j\}$ denote the time intervals between two successive discontinuity points of the failure rate (failures or pseudo-failures). The prediction of time interval $n + 1$ is:

$$x_{n+j+1;F} = \beta_{n+j+1}[\theta^{(1-F)^{-\frac{1}{n+1}}-1} - 1]; \quad \theta = \prod_{i=1}^{n+j} \left(1 + \frac{x_i}{\beta_i}\right). \tag{5.107}$$

Note that the exponent of $(1 - F)$ is not altered by the introduction of pseudo-failures, and thus it keeps track of the actual number of the interval being predicted. If it were affected, the pseudo-failures would be treated exactly like the actual failures, the system would look very unreliable because of the excessive number of failures, and the prediction would be erroneous. A pseudo-failure is inserted when the actual time-to-failure exceeds the estimate derived from previous data. Thus, the introduction of pseudo-failures is consistent with the fact that the period of failure-free working causes the reliability of DFR systems to improve. DFR is the usual for case software [11], and for hardware in the incipient period of use.

The pseudo-failures artificially enrich the experimental data, providing more time intervals for the computation of θ and a better approximation of the failure rate, by increasing the number of discontinuities of the piecewise constant failure rate, as shown in Fig. 5.11.

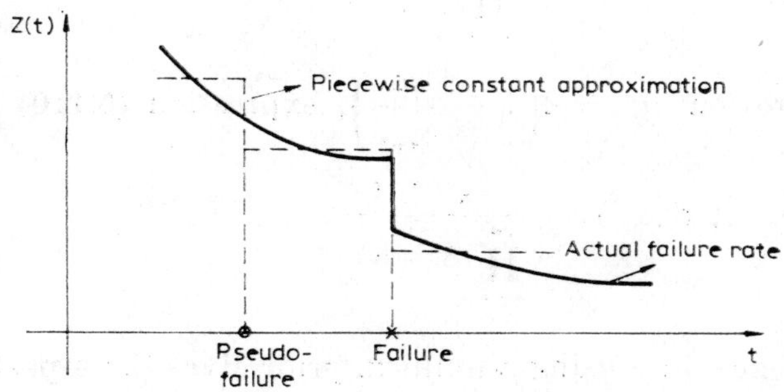

Fig. 5.11. — Approximating the failure rate using pseudo-failures.

Suppose now that an observer is situated at time instant t between failures $n - 1$ and n. It is necessary to predict the time to the nth system failure, that is, the residual life (Fig. 5.12).

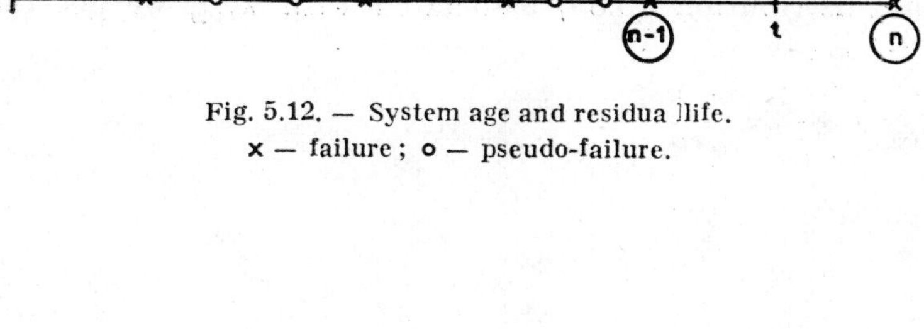

Fig. 5.12. — System age and residual life.
x — failure; o — pseudo-failure.

As the failure $n+1$ has not yet occurred, the time interval x_{n+1} does not contain pseudo-failures. The prediction instant t divides x_{n+1} into two subintervals : the system age $U(t)$ and the residual life $V(t)$. The prediction goal is the computation of interval reliability, that is $R(t, t+x) = P(V(t) > x)$. The observed data are the time intervals $x_1, x_2, \ldots, x_n$ and the system age u. As the system is piecewise exponential, the probability that it has survived from t_n to t is $e^{-\lambda u}$. Taking into account that λ_{n+1} is gamma distributed with parameters α and β_{n+1}, the total probability formula, shows that :

$$R(t_n, t/\alpha) = \int_0^\infty e^{-u\lambda} f_{a_{n+1}}(\lambda/\alpha)\,d\lambda = \frac{\beta_{n+1}^{\alpha}}{(\beta_{n+1}+u)^{\alpha}}. \tag{5.108}$$

The joint probability of the data is :

$$L(x_1, x_2, \ldots, x_n/\alpha) = \prod_{i=1}^{n} f_i(x_i/\alpha)\cdot R(t_n, t/\alpha). \tag{5.109}$$

Taking expressions (5.94) and (5.108), into account, (5.109) becomes :

$$L(x_1, x_2, \ldots, x_n, u/\alpha) = \frac{\alpha^n \theta^{-\alpha}}{\prod_{i=1}^{n}(\beta_i + x_i)} \cdot \left(\frac{\beta_{n+1}}{\beta_{n+1}+u}\right)^{\alpha}. \tag{5.110}$$

Using the notation $\theta^* = \theta\left(1 + \frac{u}{\beta_{n+1}}\right)$, expression (5.110) becomes :

$$L(x_1, x_2, \ldots, x_n, u/\alpha) = \frac{\alpha^n \theta^{*-\alpha}}{\prod_{i=1}^{n}(\beta_i + x_i)}. \tag{5.111}$$

A Bayes inference on α using a uniform prior gives the a-posteriori distribution :

$$\psi_p(\alpha) = \frac{\eta^*(\eta^*\alpha)^n e^{-\eta^*\alpha}}{\Gamma(n+1)}; \quad \eta^* = \ln\theta^*. \tag{5.112}$$

As the posterior distribution of α is of the gamma type with parameters $n+1$ and η^*, the Bayes estimate of α is the a-posteriori mean :

$$\hat{\alpha} = \frac{n+1}{\eta^*}. \tag{5.113}$$

Noting the distribution (5.89) associated with λ_{n+1}, we get the Bayes estimate of λ_{n+1} :

$$\hat{\lambda}_{n+1} = \frac{\hat{\alpha}}{\beta_{n+1}} = \frac{n+1}{\eta^*\beta_{n+1}}. \tag{5.114}$$

Similarly, a Bayes estimate of the failure rate between failure $n + k - 1$ and failure $n + k$ may readily be obtained:

$$\hat{\lambda}_{n+k} = \frac{\hat{\alpha}}{\beta_{n+k}} = \frac{n+1}{\eta^* \beta_{n+k}}. \tag{5.115}$$

We have seen that for a piecewise exponential system the residual life is exponentially distributed, so the mean residual life is:

$$\hat{m}(t) = \frac{1}{\hat{\lambda}_{n+1}} = \frac{\eta^* \beta_{n+1}}{n+1}. \tag{5.116}$$

To derive the interval reliability we use the total probability formula, taking the mean of $e^{-\lambda x}$ over the posterior distribution of α:

$$R(t, t + x) = \int_0^\infty e^{-\lambda x}\, d\lambda \int_0^\infty f_{a_{n+1}}(\lambda/\alpha)\psi_p(\alpha)\, d\alpha =$$

$$= \int_0^\infty \psi_p(\alpha)\, d\alpha \int_0^\infty e^{-\lambda x} f_{a_{n+1}}(\lambda/\alpha)\, d\lambda = \left[\frac{\eta^*}{\eta^* + \ln\left(1 + \dfrac{x}{\beta_{n+1}}\right)}\right]^{n+1}. \tag{5.117}$$

Relation (5.117) states that the system will survive from time instant t to time instant $t + x$, with probability R.

Conversely, the time-to-failure $n + 1$ such that the failure probability is F is predicted at t as:

$$x_{n+1;F} = \beta_{n+1}(\theta^{*(1-F)^{-\frac{1}{n+1}}-1} - 1). \tag{5.118}$$

Relations (5.114) and (5.118) predict the future behaviour of the system, on the basis of the a-priori information which provides the values of β_i and of the experimental information which is synthesized in the value of η^*.

The system age u may be used in a decision rule to establish whether the system reliability has attained a prescribed level. Let λ_0 be the prescribed failure rate, which may be attained by burn-in. Setting $\hat{\lambda}_{n+1} = \lambda_0$ in expression (5.114), we get:

$$\eta^* = \frac{n+1}{\lambda_0 \beta_{n+1}}. \tag{5.119}$$

As $\eta^* = \ln \theta + \ln\left(1 + \dfrac{u_i}{\beta_{n+1}}\right)$ the minimum failure-free time, which accounts for a failure rate λ_0, may be computed from relation (5.119):

$$u_{\lambda_0} = \beta_{n+1}\left[\frac{1}{\theta}\exp\left(\frac{n+1}{\lambda_0 \beta_{n+1}}\right) - 1\right]. \tag{5.120}$$

As soon as the system age has reached the value (5.120), it may be decided that the failure rate has come down to the prescribed value λ_0.

The same decision rule holds when a certain level of the mean time to failure $m_0 = 1/\lambda_0$ is prescribed.

If the requirement concerns the reliability function in time interval T, we get from expression (5.117):

$$\eta^* = \frac{\ln\left(1 + \frac{T}{\beta_{n+1}}\right)}{R^{-\frac{1}{n+1}} - 1}. \tag{5.121}$$

The minimum failure-free time which proves that the prescribed reliability R in time interval T has been reached is:

$$u_R = = \beta_{n+1}\left[\frac{1}{\theta}\left(1 + \frac{T}{\beta_{n+1}}\right)^{\left(\frac{1}{R^{-\frac{1}{n+1}} - 1}\right)} - 1\right]. \tag{5.122}$$

The model which describes the stochastically negative renewal process is a bayesian learning model. It uses a-priori information and experimental data to make predictions, which may be contradicted by future observations. These observations are subsequently used to make better predictions. The accuracy of the predictions improves as the set of experimental data becomes larger.

The assumptions on which the model is based make it suitable for describing the reliability of software systems, and of hardware systems in the incipient period of their use.

5.4 Availability

In the analysis of repairable systems given in the preceding Sections, we neglected the time durations necessary to complete the maintenance operations.

Reversing this omission, we notice that the renewal operation duration depends on the intrinsic ease of maintenance of the system (rapid fault diagnosis and fault correction) and on the organization of system maintenance (logistics, testing equipment etc.).

In order to describe, mathematically, the ease of maintenance (maintainability) of a system, we shall consider the down time Y, that is, the random time interval starting from the instant of failure to the instant when the repair is completed. It is assumed that the system is restored by repair into the same state, whether it is the initial state or not. The duration of the renewal is analogous to the time-to-failure of the system, as both are continuous random variables. The maintainability model may then be readily derived from the global reliability model, by appropriately re-interpreting the reliability measures defined in Chapter 2.

For instance, $G(t)$, the cdf of the down time, is defined as the probability that the repair is completed in $(0, t)$. The pdf $g(t)$ of the down time, multiplied by Δt, is the total probability that the repair is completed around time instant t. The repair rate $\rho(t)$, multiplied by Δt, is the conditional probability that the repair is completed around time instant t, given that it was not completed up to t. Table 5.1 provides a complete list of maintainability measures, with their mathematical definitions and intuitive interpretations. A satisfactory maintainability implies high values for $G(t)$, $g(t)$ and $\rho(t)$, and low values for μ, σ_Y and y_F.

Table 5.1

Maintainability measures

Symbol	Definition	Interpretation
$G(t)$	$G(t) = P(Y < t)$	Probability that the repair is completed in $(0, t)$
$g(t)$	$g(t) = G'(t)$	$g(t)\Delta t$ is the probability that the repair is completed around time instant t
$\rho(t)$	$\rho(t) = \dfrac{g(t)}{1-G(t)}$	$\rho(t)\,\Delta t$ is the conditional probability that the repair is completed around t given that it was not completed before t
μ	$\mu = \int_0^\infty t g(t)\,dt$	Mean time to repair (MTTR)
V_Y	$V_Y = (t-\mu)^2 g(t)dt$	Variance of down time
σ_Y	$\sigma_Y = \sqrt{V}$	Standard deviation of down time
y_F	$G(y_F) = F$	F-percentile of down time

Knowledge of the reliability and maintainability measures allows a description of the renewal process. In this process, the active periods $X_1, X_2, \ldots$ alternate with the repair durations $Y_1, Y_2, \ldots$. (Fig. 5.13). In the simple alternating process, $X_1, X_2, \ldots, X_r, \ldots$, are iid random

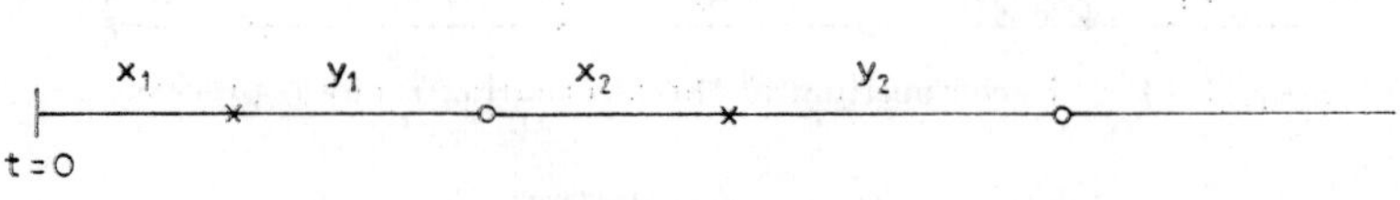

Fig. 5.13. — Alternating renewal process.

variables distributed according to the same law $f(t)$ and $Y_1, Y_2, \ldots, Y_r, \ldots$ are iid random variables distributed according to $g(t)$. The simple alternating renewal process describes the situation when the system recovers, after failure, to its initial state.

In the general alternating renewal process, the first time-to-failure is distributed according to $f_1(t)$, while all the other times-to-failure (x_2, x_3, ...) are distributed according to $f(t)$. This is the case when the system recovers, after each failure, into the same state but one which is different from the initial state.

The system failures and recoveries form two separate renewal processes. The failure process is called a type I renewal process and is a general renewal process with negligible renewal durations. The process of recovery is called a type II renewal process and may be a simple or a general renewal process with negligible renewal durations. The alternating renewal process is a superposition of the two renewal processes described above (Fig. 5.14). It is straightforward to analyse the type I and the type II renewal processes separately.

In the type II renewal processes, the time intervals between renewals are $X_1 + Y_1$, $X_2 + Y_2$, ..., $X_r + Y_r$, ... If the alternating process is simple, all these variables are iid according to the convolution $f(t) \otimes g(t)$. If the alternating process is general, the first cycle $X_1 + Y_1$ is distributed according to $f_1(t) \otimes g(t)$, while all other cycles are distributed according to $f(t) \otimes g(t)$. The Laplace transform of the type II renewal density may then be derived from relations (5.32) and (5.33). In the case of the simple process, we get:

$$h_2^*(s) = \frac{f^*(s)g^*(s)}{1 - f^*(s)g^*(s)}. \tag{5.123}$$

In the case of the general process, we get

$$h_2^*(s) = \frac{f_1^*(s)g^*(s)}{1 - f^*(s)g^*(s)}. \tag{5.124}$$

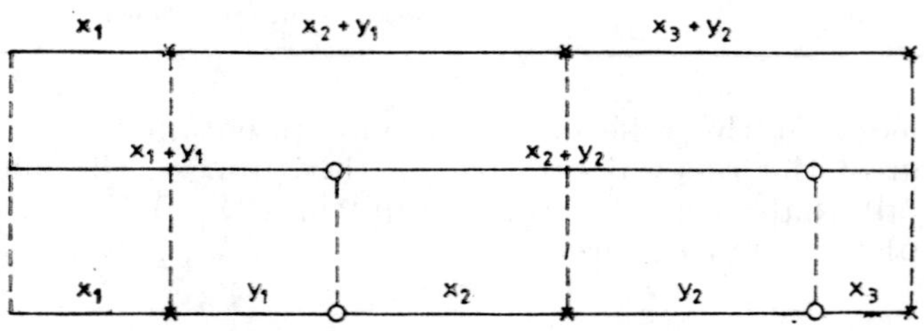

Fig. 5.14. — Decomposition of the alternating renewal process.

The type I renewal process is a general process, even if the system recovers into its initial state. The time intervals between type I renewals are X_1, $Y_1 + X_2$, $Y_2 + X_3$, ... The first interval is distributed according to $f(t)$ in a simple process and according to $f_1(t)$ in a general process, while all the other intervals are distributed according to the convolution $f(t) \otimes \otimes g(t)$. Using relations (5.32) and (5.33), we get the type I renewal density for the case of simple and general processes respectively :

$$h_1^*(s) = \frac{f^*(s)}{1 - f^*(s)g^*(s)}; \quad h_1^*(s) = \frac{f_1^*(s)}{1 - f^*(s)g^*(s)}. \tag{5.125}$$

Both type I and type II renewal processes are asymptotically stationary, so :

$$\lim_{t\to\infty} h_1(t) = \lim_{t\to\infty} h_2(t) = \frac{1}{m + \mu}. \tag{5.126}$$

Once the renewal densities are known, it is easy to derive the renewal functions, that is, the mean number of failures and the mean number of recoveries in a specified time interval. In the stationary state, the mean number of failures and the mean number of recoveries in $(t, t + x)$ are both equal to the time interval x divided by the mean of the repetitive cycle $m + \mu$.

The renewal functions allow a reasonable planning of resources for system maintenance.

The future system behaviour may be predicted by deriving the interval reliability $R(t, t + x)$. Let us consider the time instant t which divides an active period into two parts : the system age $U(t)$ and the residual life $V(t)$ (Fig. 5.15). The residual life $V(t)$ is greater than x in the following situations :

a) The system does not fail up to $t + x$; this event has probability $R(t + x)$ in the case of a simple process, and $R_1(t + x)$ in the case of a general process.

b) The system recovers around $t - u$ and does not fail up to $t + x$. This event has the probability

$$h_i(t - u)R(u + x)\,\mathrm{d}u.$$

Taking the situations (a) and (b) into account, we find by the total probability formula :

$$R(t, t + x) = R(t + x) + \int_0^t h_2(t - u)R(u + x)\,\mathrm{d}u \tag{5.127}$$

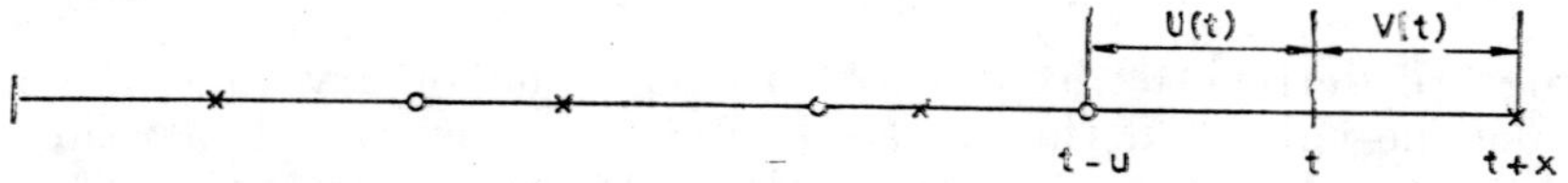

Fig. 5.15. — Prediction of system behaviour.

in the case of a simple process, and

$$R(t, t + x) = R_1(t + x) + \int_0^t h_2(t - u)R(u + x)\,\mathrm{d}u \tag{5.128}$$

in the case of a general process.

It should be noted that the reliability function in a time interval starting from the instant of a recovery depends only on the intrinsic reliability characteristics of the system. The reliability function in $(t, t + x)$ depends not only on the intrinsic reliability but also on the system maintainability and on the maintenance procedures.

In the stationary state, the interval reliability function is :

$$\lim_{t\to\infty} R(t, t + x) = \frac{\int_x^\infty R(t)\,\mathrm{d}t}{m + \mu}. \tag{5.129}$$

The mean residual life is obtained by integrating the interval reliability function with respect to x :

$$m(t) = \int_0^\infty R(t, t + x)\,\mathrm{d}x. \tag{5.130}$$

In the stationary state, relation (5.130) becomes :

$$\lim_{t\to\infty} m(t) = \frac{1}{m + \mu}\int_0^\infty \mathrm{d}x \int_x^\infty R(t)\,\mathrm{d}t = \frac{1}{m + \mu}\int_0^\infty xR(x)\,\mathrm{d}x. \tag{5.131}$$

The interval reliability and the mean residual life are useful for predicting the system behaviour.

Some specific reliability measure is needed to describe the overall system effectiveness. The effectiveness is related to the total active operational time $\Gamma(t)$ of the system in the calendar time interval $(0, t)$. The mathematical description of the random variable $\Gamma(t)$ based on $f(t)$ and $g(t)$ is rather complicated [12, 13, 14], so only its average value will be considered. The mean active time divided by the calendar time is called the interval availability :

$$A(t) = \frac{1}{t}\, E_{\Gamma(t)}. \tag{5.132}$$

In order to derive the expression $\tilde{A}(t)$, let the binary random variable $S(t)$ take the value 1 if the system is up (i.e., operational) at time t and 0 in the opposite case. The total active time may be written:

$$\Gamma(t) = \int_0^t S(t)\,\mathrm{d}t. \tag{5.133}$$

The probability that the system is up at t is called the availability :

$$A(t) = P[S(t) = 1] = E_{S(t)}. \tag{5.134}$$

Taking the mean of $\Gamma(t)$, we get :

$$\tilde{A}(t) = \frac{1}{t} E_{\Gamma(t)} = \frac{1}{t}\int_0^t E_{S(t)}\,dt = \frac{1}{t}\int_0^t A(t)\,dt. \tag{5.135}$$

Thus, knowledge of the system availability $A(t)$ implies the possibility of computing the active fraction of the calendar time. Availability is a synthetical measure of renewal systems reliability; it results from combining of intrinsic reliability, maintainability and maintenance procedures.

The availability equation may be readily obtained by setting $x = 0$ in expression (5.127) or (5.128). We get

$$A(t) = R(t) + \int_0^t h_2(t-u)\,R(u)\,du$$

in the simple process, and

$$A(t) = R_1(t) + \int_0^t h_2(t-u)\,R(u)\,du \tag{5.136}$$

in the general process.

Equations (5.136) may be interpreted as follows. The system is up at t, either if it has not failed in $(0, t)$, or if it recovered around $t-u$ $(u \in (0, t))$ and has not failed since.

Taking the Laplace transform of equations (5.136), we find in the case of the generalized process :

$$A^*(s) = R_1^*(s) + h_2^*(s)\,R^*(s) = \frac{1-f_1^*(s)}{s} +$$

$$+ \frac{f_1^*(s)g^*(s)}{1-f^*(s)\,g^*(s)} \cdot \frac{1-f^*(s)}{(s)} = H_2^*(s) - H_1^*(s) + \frac{1}{s} \cdot \tag{5.137}$$

Relation (5.137) is equivalent to :

$$\bar{A}(t) = 1 - A(t) = H_1(t) - H_2(t). \tag{5.138}$$

The system unavailability at t is equal to the difference between the mean number of failures and the mean number of recoveries in $(0, t)$. This result could have been obtained directly. The number of failures is equal to the number of recoveries if t belongs to an active time interval, and is greater by one than the number of recoveries if t belongs to a repair time interval, so that :

$$\begin{aligned} N_{1t} - N_{2t} &= 0\,; \quad t \in X_i \\ N_{1t} - N_{2t} &= 1\,; \quad t \in Y_i. \end{aligned} \tag{5.139}$$

Taking the mean of $N_{1t} - N_{2t}$, we get relation (5.138).

The asymptotic expression of availability may be obtained by letting $t \to \infty$ in relations (5.136).

$$\lim_{t\to\infty} A(t) = A = \frac{1}{m+\mu}\int_0^{\infty} R(t)\,\mathrm{d}t = \frac{m}{m+\mu}. \tag{5.140}$$

Expression (5.140) shows that, in the stationary state, the probability that the system is in a working state at t is constant and equal to the active fraction of an average repetitive cycle. From the stationary expressions of the interval reliability and of the availability (5.129) and (5.140), we get :

$$\lim_{t\to\infty} R(t, t+x) = \frac{m}{m+\mu}\cdot\frac{1}{m}\int_x^{\infty} R(t)\,\mathrm{d}t = AR(x/t). \tag{5.141}$$

The interval reliability is then a total probability, equal to the availability multiplied by the interval reliability conditioned by the system functioning at the beginning of the time interval. For CFR systems, the conditioned interval reliability is :

$$R(x/t) = \frac{1}{m}\int_x^{\infty} R(t)\,\mathrm{d}t = \mathrm{e}^{-\lambda x}. \tag{5.142}$$

The interval reliability of a system without memory is then the product of the availability and the reliability function. This result holds also in the transitory state of the process.

We shall now show that the interval availability $\tilde{A}(t)$ has the same asymptotic value A as the availability. Let us consider the value t_ε, such that for any $t > t_\varepsilon$ we have

$$|A(t) - A| < \varepsilon. \tag{5.143}$$

The interval availability $\tilde{A}(t)$ is, for $t > t_\varepsilon$:

$$\tilde{A}(t) = \frac{1}{t}\int_0^{t} A(t)\,\mathrm{d}t = \frac{1}{t}\int_0^{t_\varepsilon} A(t)\,\mathrm{d}t + \frac{1}{t}\int_{t_\varepsilon}^{t} A(t)\,\mathrm{d}t. \tag{5.144}$$

Taking into account relation (5.143), we derive from (5.144) the inequalities :

$$\frac{1}{t}\int_0^{t_\varepsilon} A(t)\,\mathrm{d}t + \frac{1}{t}\int_{t_\varepsilon}^{t}(A-\varepsilon)\mathrm{d}t \leqslant \tilde{A}(t) \leqslant \frac{1}{t}\int_0^{t_\varepsilon} A(t)\,\mathrm{d}t + \\ + \frac{1}{t}\int_{t_\varepsilon}^{t}(A+\varepsilon)\,\mathrm{d}t. \tag{5.145}$$

For t sufficiently great, inequalities (5.145) become:

$$A - \varepsilon \leqslant \tilde{A}(t) \leqslant A + \varepsilon \tag{5.146}$$

that is, $\lim_{t\to\infty} \tilde{A}(t) = A$.

This result may be used to compute the active mean fraction of a long time span, by dividing the mean time-to-failure by the mean duration of a repetitive cycle.

We have shown that the availability may be derived from the reliability function and the maintainability measure. Sometimes it is easier to find the availability by direct methods [15] and to derive the reliability function from the availability expression. To do this, we start by differentiating expression (5.138):

$$\frac{\mathrm{d}A(t)}{\mathrm{d}t} = -h_1(t) + h_2(t). \tag{5.147}$$

We define the function $\Lambda(t) = h_1(t)/A(t)$, so that $\Lambda(t)\Delta t$ represents the conditioned probability of failure in $(t, t + \Delta t)$, given that the system is up at time t. Relation (5.147) becomes:

$$\frac{\mathrm{d}A(t)}{\mathrm{d}t} = -\Lambda(t)\, A(t) + h_2(t). \tag{5.148}$$

Integrating equation (5.148) with initial condition $A(0) = 1$:

$$A(t) = \exp\left[-\int_0^t \Lambda(u)\,\mathrm{d}u\right] + \int_0^t h_2(t-u)\exp\left[-\int_0^u \Lambda(\tau)\,\mathrm{d}\tau\right]\mathrm{d}u. \tag{5.149}$$

Comparing relations (5.136) and (5.149) for a simple alternating renewal process, we derive

$$R(t) = \exp\left[-\int_0^t \Lambda(u)\,\mathrm{d}u\right]. \tag{5.150}$$

Relation (5.150) contradicts the general expression of the reliability function:

$$R(t) = \exp\left[-\int_0^t z(u)\,\mathrm{d}u\right]. \tag{5.151}$$

The functions $\Lambda(t)$ and $z(t)$ are identical only for non-repairable systems and for repairable CFR (memoryless) systems. In all other cases they are different, because $\Lambda(t)\Delta t$ is a failure probability conditioned by the system being up at t while $z(t)\Delta t$ is a failure probability conditioned by the system acting satisfactorily in $(0, t)$. It follows that relation (5.150) is incorrect, as it was derived from relation (5.138) which refers to average values. It is not permissible to deduce the equality of two functions from the equality of their mean values.

Nevertheless, relation (5.150) is useful as an approximation, as shown by the following inequalities derived in [15]:

$$\begin{aligned} &H_1(t) \geqslant F(t) \geqslant \bar{A}(t) \\ &H_1(t) \geqslant 1 - \exp\left[-\int_0^t \Lambda(u)\,\mathrm{d}u\right] \geqslant \bar{A}(t). \end{aligned} \tag{5.152}$$

It is also shown [15] that the approximation is better when the failure rate is more uniform.

We shall subsequently use the renewal reliability models described in this Section in order to develop the maintenance procedures able to improve system reliability and efficiency.

5.5 Renewal policies

The system effectiveness may be improved by preventive renewal of the system prior to failure. The preventive renewal eliminates the cumulated wear (it is of the "good as new" type) and thus prolongs the time to failure of the system. It is clear that preventive renewal is efficient only for IFR and IFRA systems.

A renewal policy is specified by the scheduled times of preventive renewals. Renewals in the event of system failure are carried out at random times, whereas preventive renewals may be deterministic or random events, according to the type of renewal policy. The preventive renewals form another renewal process, superposed on the original renewal process generated by the system failures.

There are two main types of renewal policies: periodic and non-periodic. The periodic renewal policy is deterministic, as it is characterized by a constant time interval between any two successive preventive renewals. The non-periodic renewal policies are based on system age or on system wear, so they are random renewal processes.

Several criteria may be used in formulating renewal policies.

A renewal policy may be designed so that operational reliability should exceed a specified lower bound. Another criterion is that the renewal policy should maximize the system availability.

Finally, the renewal policy may be designed so that the average maintenance cost rate should be minimized. It is important to evaluate the maintenance cost, even if the renewal policy is based on the criteria of maximum availability or specified operational reliability. To this end, the cost of a renewal upon failure is taken as unity and the costs of preventive renewals are expressed by fractions of the cost of a renewal upon failure.

If the above criteria are used separately in the design of a specific renewal policy, different scheduled times for preventive renewal will be obtained. Multiple-criteria decision-making is used to select the optimal renewal policy [16]. The method requires that the criteria be ranked in order of importance by the decision-maker. The preferred solution is the one which optimizes the criteria, starting with the most important and proceeding according to the order of importance of the criteria. An alternative procedure is to optimize the first criterion, then to optimize the second criterion subject to keeping the first criterion within a certain percentage of its optimum. The third criterion is then optimized keeping the first two within a certain percentage of the optimum values found in the previous step, and so on.

We shall consider the simplest periodic renewal policy, the block replacement policy (BRP). According to this, the system is restored to its initial state either upon failure or at equally spaced time instants $kT\ (k = 1, 2, \ldots)$.

The time interval T between two successive preventive renewals may be determined so that the reliability function of an operational run which starts at kT and lasts until $(k + 1)T$ should not be lower than a specified level R_0:

$$R(T) \geqslant R_0. \tag{5.153}$$

In order to compute the average maintenance cost rate, we shall assume that all renewals are instantaneous.

The average maintenance cost of the system between two successive preventive renewals includes the cost b of a preventive renewal and the cost of the renewals upon failure which occur in $[kT, (k + 1)T]$. As the mean number of failures in such a time interval is $H(T)$, the average maintenance cost rate is

$$k_{\mathrm{BRP}} = \frac{b + H(T)}{T}. \tag{5.154}$$

This cost must be compared to the average maintenance cost rate when no preventive renewals are carried out. We shall describe this situation as a failure replacement policy (FRP). The average cost rate in the stationary state of the renewal process is in this case

$$k_{\mathrm{FRP}} = \frac{1}{m}. \tag{5.155}$$

We shall consider, as an example, a parallel two-unit system which has the reliability function

$$R(t) = 2\mathrm{e}^{-\lambda t} - \mathrm{e}^{-2\lambda t}; \quad \lambda = 10^{-4}\mathrm{h}^{-1}. \tag{5.156}$$

Such a system fails when both its units fail. Its mean time to failure is $m = \int_0^\infty R(t)\,\mathrm{d}t = \frac{3}{2\lambda}$. The block replacement policy is based on the criterion that the reliability function in the time interval between two successive preventive renewals should not be less than $R_0 = 0.99$. From relations (5.153) and (5.156) we get:

$$\begin{aligned} &\mathrm{e}^{-2\lambda T} - 2\mathrm{e}^{-\lambda T} + 0.99 \leqslant 0\,; \\ &T \leqslant \frac{1}{\lambda} \ln \frac{11}{0.9} = 1053.6 \text{ h}. \end{aligned} \tag{5.157}$$

In order to find the average maintenance cost rate, the renewal function must first be computed, using the method described in Section 5.1. Its successive steps are as follows:

$$\begin{aligned} f(t) &= -\frac{\mathrm{d}R(t)}{\mathrm{d}t} = 2\lambda\mathrm{e}^{-\lambda t} - 2\lambda\mathrm{e}^{-2\lambda t}\,; \\ f^*(s) &= \frac{2}{s+\lambda} - \frac{2}{s+2\lambda} = \frac{2\lambda^2}{(s+\lambda)(s+2\lambda)}\,; \\ h^*(s) &= \frac{f^*(s)}{1-f^*(s)} = \frac{2\lambda^2}{s(s+3\lambda)}\,; \\ h(t) &= \frac{2\lambda}{3}(1-\mathrm{e}^{-3\lambda t})\,; \\ H(t) &= \int_0^t h(t)\,\mathrm{d}t = \frac{2\lambda t}{3} + \frac{2}{9}\mathrm{e}^{-3\lambda t} - \frac{2}{9}. \end{aligned} \tag{5.158}$$

Suppose that the cost of a preventive renewal is 20% of the cost of a renewal upon failure ($b = 0.2$). By relations (5.154) and (5.157), the average maintenance cost rate is:

$$k_{\mathrm{BRP}} = \frac{\frac{2\lambda T}{3} + \frac{2}{9}\mathrm{e}^{-3\lambda T} - \frac{2}{9} + 0.2}{T} = 3.08 \times 10^{-4}\,\mathrm{h}^{-1}. \tag{5.159}$$

In a failure replacement policy, the average maintenance cost rate according to relation (5.155) is:

$$k_{\mathrm{FRP}} = \frac{1}{m} = \frac{2\lambda}{3} = 0.66 \times 10^{-4}\,\mathrm{h}^{-1}. \tag{5.160}$$

Thus the average maintenance cost of the block replacement policy is higher, but the reliability function in the BRP case does not fall below a specified level. From the point of view of the maintenance cost, the block replacement policy is never economical in the stationary state of the renewal process. We have, indeed :

$$k_{\mathrm{BRP}} = \frac{1}{m} + \frac{b}{T} > \frac{1}{m} = k_{\mathrm{FRP}}. \tag{5.161}$$

The block replacement policy would reduce the maintenance cost only during the transitory state of the renewal process.

We shall now use, as a criterion for developing a block replacement policy, minimization of the average maintenance cost rate. The equation for the optimal time interval T^* between preventive renewals is :

$$\begin{aligned} &\frac{\partial k_{\mathrm{BRP}}}{\partial T} = 0 ; \\ &T^* h(T^*) - H(T^*) = b. \end{aligned} \tag{5.162}$$

The solution of equation (5.162) exists if the renewal density $h(t)$ is an increasing function of t[17]. The minimum average maintenance cost rate is :

$$k^*_{\mathrm{BRP}} = \frac{H(T^*) + b}{T^*} = h(T^*). \tag{5.163}$$

k^*_{BRP} is an increasing function of b, by the intermediary of T^*.

In the example considered above, the equation of the optimal time interval between preventive renewals becomes :

$$\left(1 - \frac{9}{2} b\right) \mathrm{e}^{y^*} = y^* + 1 \text{ wehere } y^* = 3\lambda T^*. \tag{5.164}$$

The solution exists only for $b < 2/9$. Equation (5.164) is solved graphically or by computer for $b = 0.2$ and $\lambda = 10^{-4}\mathrm{h}^{-1}$, yielding T^*.

The optimal average maintenance cost rate is :

$$\begin{aligned} k^*_{\mathrm{BRP}} &= h(T^*) = \frac{2\lambda}{3}(1 - \mathrm{e}^{-y^*}) = \frac{2\lambda}{3} \frac{y^* + 0.9}{y^* + 1} < \\ &< \frac{2\lambda}{3} = k_{\mathrm{FRP}}. \end{aligned} \tag{5.165}$$

The difficulty in the design of block replacement policies lies in the impossibility of computing the Laplace transforms, for some life distributions. This applies in the case of the Weibull distribution for which a special method to derive an optimal BRP has been developed [18].

The average maintenance cost rate k_{BRP} is situated between two cost rates; they are k_{B_0}, the average cost rate of BRP when only one failure occurs between two successive preventive renewals, and k_{B_2}, the average cost rate when minimal (bad as old) repairs are performed upon system failure. It may be shown [18] that:

$$k_{B_0} < k_{\mathrm{BRP}} < k_{B_2}. \tag{5.166}$$

It is easy to derive the cost rates k_{B_0} and k_{B_2}:

$$k_{B_0} = \frac{F(T) + b}{T}; \quad k_{B_2} = \frac{\int_0^T z(t)\,\mathrm{d}t + b}{T}. \tag{5.167}$$

Setting the first derivatives of expressions (5.167) equal to zero we get two different time intervals T_0^* and T_2^* which minimize k_{B_0} and k_{B_2} respectively:

$$\begin{aligned} &T_0^* f(T_0^*) - F(T_0^*) = b; \\ &T_0^* z(T_2^*) - \int_0^{T_2^*} z(t)\,\mathrm{d}t = b. \end{aligned} \tag{5.168}$$

T_0^* and T_2^* may easily be computed from equations (5.168) and the optimal T^* which minimizes k_{BRP} is approximated by a value between T_0^* and T_2^*.

A strict periodic replacement policy has the disadvantage of inflexible planning, so that the time for preventive renewal could come shortly after a renewal upon failure has been carried out. To prevent such a situation, a modified block replacement policy has been developed [19]. Preventive renewals in this scheme are also periodic and are performed at time instants kT $(k = 1, 2, \ldots)$, but when a failure occurs in time interval $(kT - T_D, kT)$ the repair is delayed until the time instant kT of the next planned renewal. This modified block replacement policy is called DRP (delayed replacement policy). The maintenance cost includes in this case the cost d of the system being down (non-operational) or acting incorrectly during a time unit. The average maintenance cost in $((k-1)T, kT)$ has the following components:

a) cost of preventive renewal b;

b) cost of the average number of renewals upon failure in $((k-1)T, kT - T_D)$, which is $H(T - T_D)$;

c) cost of mean down time $d \int_0^{T_D} x h(t - x)\,\mathrm{d}x$.

The average maintenance cost rate is then:

$$k_{\mathrm{DRP}}(T, T_D) = \frac{1}{T}\left[b + H(T - T_D) + d\int_0^{T_D} xh(T - x)\,\mathrm{d}x\right]. \tag{5.169}$$

Minimizing $k_{\mathrm{DRP}}(T, T_D)$ with respect to T_D yields the optimal value:

$$T_D^* = \frac{1}{d}. \tag{5.170}$$

For $T_D^* \ll T$, expression (5.169) becomes:

$$k_{\mathrm{DRP}}(T, T_D^*) \approx k_{\mathrm{BRP}}(T) - \frac{h(T)}{2dT}. \tag{5.171}$$

Replacing now T by the optimal T^* given by equation (5.162) we get:

$$k_{\mathrm{DRP}}^* = k_{\mathrm{DRP}}(T^*, T_D^*) = h(T^*)\left(1 - \frac{1}{2dT^*}\right) =$$

$$= h(T^*)\left(1 - \frac{T_D^*}{2T^*}\right). \tag{5.172}$$

A block replacement policy may be designed so that the availability coefficient should meet some specified requirement. If availability is used as a design criterion, the renewal durations cannot be neglected. Let μ_1 and μ_2 be the mean durations of a renewal upon failure and of a preventive renewal, respectively. The mean down time between two successive preventive renewals is:

$$T_S = \mu_1 H_1((k - 1)T, kT) + \mu_2 \tag{5.173}$$

where $H_1((k - 1)T, kT)$ is the type I renewal function in time interval $((k - 1)T, kT)$.

In the stationary state, we have

$$H_1((k - 1)T, kT) = \frac{T}{m + \mu_1}. \tag{5.174}$$

The mean down time is then:

$$T_S = \frac{\mu_1 T}{m + \mu_1} + \mu_2. \tag{5.175}$$

The availability coefficient under BRP is the ratio between the mean active time and the period of the preventive renewals:

$$A_{\mathrm{BRP}} = \frac{T - T_S}{T} = \frac{m}{m + \mu_1} - \frac{\mu_2}{T} = A_{\mathrm{FRP}} - \frac{\mu_2}{T} < A_{\mathrm{FRP}}. \tag{5.176}$$

It follows from relation (5.176) that the block replacement policy cannot improve the availability coefficient, relative to the FRP case, when no preventive renewals are carried out. This conclusion is in agreement with the fact that a block replacement policy is not economic in the stationary state of the renewal process (see relation (5.161)).

We shall now consider the age replacement policy (ARP), in which preventive renewal is performed if and only if the system has reached a specified age x. As this is a non-periodic policy, the cost associated with a preventive renewal should be higher than in the case of the block replacement policy ($a > b$). The criteria used to design an age replacement policy are the same as those considered in the BRP case. If the criterion of a minimum reliability level is employed, a replacement age x results equal to the period T given by equation (5.153) for the BRP case.

We consider next the criterion of a specified availability coefficient. As the system is replaced at age x, its lifetime is distributed according to:

$$\psi(t) = \begin{cases} f(t) & t < x \\ R(t) & t = x \\ 0 & t > x. \end{cases} \tag{5.177}$$

The mean lifetime of the system under ARP is:

$$m_a = \int_0^x t f(t)\,\mathrm{d}t + xR(x) = \int_0^x R(t)\,\mathrm{d}t. \tag{5.178}$$

The mean renewal duration is μ_1 if failure occurs (with probability $F(x)$) before the system reaches age x, and μ_2 otherwise. The availability coefficient is the ratio between the active mean time and the total mean time of a cycle:

$$A_{\mathrm{ARP}} = \frac{m_a}{m_a + \mu_1 F(x) + \mu_2 R(x)} = \frac{\int_0^x R(t)\,\mathrm{d}t}{\int_0^x R(t)\,\mathrm{d}t + \mu_1 - (\mu_1 - \mu_2)R(x)}. \tag{5.179}$$

When no preventive renewals are performed, the availability coefficient is:

$$A_{\mathrm{FRP}} = \frac{m}{m + \mu_1} = \frac{\int_0^\infty R(t)\,\mathrm{d}t}{\int_0^\infty R(t)\mathrm{d}t + \mu_1}. \tag{5.180}$$

Comparing relations (5.179) and (5.180), we conclude that an age replacement policy can improve the availability coefficient with respect to the FRP case if and only if:

$$\frac{\int_0^x R(t)\,\mathrm{d}t}{\int_0^\infty R(t)\,\mathrm{d}t} > 1 - R(x)\left(1 - \frac{\mu_2}{\mu_1}\right). \tag{5.181}$$

Condition (5.181) is fulfilled if $\mu_2 < \mu_1$ and the failure rate of the system is a strictly increasing function [20]. The condition is also fulfilled if the failure rate increases asymptotically to an upper limit l which satisfies the inequality [20]:

$$\frac{1}{l} < m\left(1 - \frac{\mu_2}{\mu_1}\right). \tag{5.182}$$

In these cases, the age at which preventive renewal should be performed may be computed either by specifying the value of the availability coefficient (5.179) or by maximizing expression (5.179).

An age replacement policy may also be designed so that the average maintenance cost rate should be minimized. In this case the renewal durations are neglected. The maintenance cost of the system in its lifetime is considered equal to unity if the system fails before it reaches age x (with probability $F(x)$), and equal to a in the opposite case. Dividing the average maintenance cost by the mean lifetime we get:

$$k_{\mathrm{ARP}} = \frac{F(x) + aR(x)}{\int_0^x R(t)\,\mathrm{d}t.}. \tag{5.183}$$

Minimizing expression (5.183), we get the equation of the optimal age x^* when preventive renewal should be carried out:

$$\frac{\partial k_{\mathrm{ARP}}(x)}{\partial x} = 0$$

$$z(x^*)\int_0^{x^*} R(t)\,\mathrm{d}t + R(x^*) = \frac{1}{1-a}. \tag{5.184}$$

A solution of equation (5.184) exists if the failure rate is an increasing function of time [20]. The minimum average maintenance cost rate is:

$$k^*_{\mathrm{ARP}} = k_{\mathrm{ARP}}(x^*) = (1 - a)\,z(x^*). \tag{5.185}$$

The minimum average maintenance cost rate is an increasing function of a. The optimal age x^* may also be computed by a graphical method which uses the Barlow-Campo function $t_s(F) = (1/m) \int_0^{t_F} R(t)\,dt$ [21]. Let us consider the reciprocal of the average maintenance cost rate under ARP :

$$\frac{1}{k_{\text{ARP}}} = \frac{\int_0^x R(t)\,\mathrm{d}t}{F(x) + aR(x)} = \frac{\int_0^x R(t)\,\mathrm{d}t}{(1-a)F(x) + a}. \tag{5.186}$$

Age x is the F percentile of the time-to-failure, so that expression (5.186) may be written :

$$\frac{1}{k_{\text{ARP}}} = \frac{m}{a} \frac{t_S(F)}{\frac{a}{1-a} + F}. \tag{5.187}$$

We obtain first the optimal F^* which maximizes expression (5.187)

$$\frac{\partial}{\partial F}\left(\frac{1}{k_{\text{ARP}}}\right) = 0\,;$$

$$\frac{\mathrm{d}t_S(F)}{\mathrm{d}F} = \frac{t_S(F)}{\frac{a}{1-a} + F}. \tag{5.188}$$

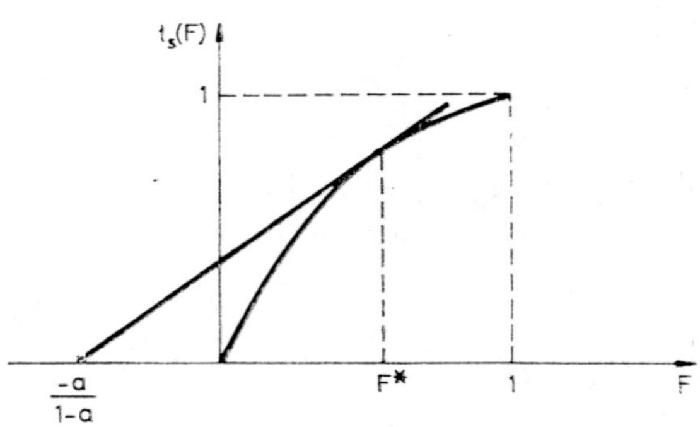

Fig. 5.16. — Design of the age replacement policy.

Equation (5.188) may be solved graphically (Fig. 5.16). From the point $\left(-\frac{a}{1-a}, 0\right)$ we draw a tangent to the graph of $t_S(F)$. The tangent point gives the optimal F^*, and hence the optimal replacement age may be computed as $F(x^*) = F^*$. The solution exists only for IFR systems, because only for such systems is the function $t_S(F)$ concave. Moreover, the solution exists for IFR systems for any value $a < 1$ if the derivative $\mathrm{d}t_S(F)/\mathrm{d}F$ at $F = 1$ is zero, which implies that the failure rate tends to infinity. If the failure rate tends asymptotically to a constant l, then the derivative $\mathrm{d}t_S(F)/\mathrm{d}F$ at $F = 1$ is $1/ml$. The tangent at (1,1) to $t_S(F)$ intersects the horizontal axis at $1 - ml$. However, if $-a/(1-a)$ is less than $(1 - ml)$, no tangent may be drawn and a solution of equation (5.188) does not exist. If $(1 - ml) = -a/(1-a)$ the solution is $F^* = 1$, that is, no preventive maintenance should be done. A unique solution exists if $(1 - ml) < -a/(1-a)$. Taking the parallel system as an example, equation (5.184) becomes :

$$(1-a)R^{*2} - (2-3a)R^* + 1 - 3a = 0 \text{ where } R^* = \mathrm{e}^{-\lambda x^*}. \quad (5.189)$$

Equation (5.189) has a solution in (0,1) for $a < 1/3$. The solution is :

$$R^* = \frac{2 - 3a - \sqrt{a(4-3a)}}{2(1-a)};$$

$$x^* = \frac{1}{\lambda} \ln \frac{2(1-a)}{2 - 3a - \sqrt{a(4-3a)}}. \quad (5.190)$$

The minimum average maintenance cost rate is :

$$k^*_{\mathrm{ARP}} = (1-a)\, z(x^*) = 2\lambda(1-a)\, \frac{a + \sqrt{a(4-3a)}}{2 - a + \sqrt{a(4-3a)}}. \quad (5.191)$$

For $a = 25\%$ and $\lambda = 10^{-4}\,\mathrm{h}^{-1}$, we get

$$x^* = 14\,552 \text{ h}; \quad k^*_{\mathrm{ARP}} = 0.65 \times 10^{-4}\,\mathrm{h}^{-1}. \quad (5.192)$$

In a failure replacement policy the average maintenance cost rate is $0.66 \times 10^{-4}\,\mathrm{h}^{-1}$, so the optimal age replacement policy is economically advantageous.

The choice between ARP, BRP and FRP is based on a comparison of their properties for the same design criterion. If the design criterion is minimization of the average maintenance cost rate, the choice will depend on comparison of the minimal cost rates in the optimal versions of the renewal policies.

It is obvious that the minimal cost rates k^*_{ARP} and k^*_{BRP} are increasing functions of the costs of preventive renewals a and b. Hence, there are two unique values a_0 and b_0 such that for any $a > a_0$ and $b > b_0$ we have :

$$k^*_{\mathrm{ARP}}(a) > k_{\mathrm{FRP}}; \; a > a_0$$

$$k^*_{\mathrm{BRP}}(b) > k_{\mathrm{FRP}}; \; b > b_0. \quad (5.193)$$

From inequalities (5.193) it follows that for $a > a_0$ and $b > b_0$ a failure replacement policy must be adopted, for $a > a_0$ and $b < b_0$ we must adopt a block replacement policy while for $a < a_0$ and $b > b_0$ an age replacement policy is preferable.

In the case $a < a_0$ and $b < b_0$, the minimal cost rates k^*_{ARP} and k^*_{BRP} should be compared. As both are increasing functions, of a and b respectively, there is a unique value $b^*(a)$ for any given a such that $k^*_{\mathrm{BRP}}(b^*) = k^*_{\mathrm{ARP}}(a)$. It is obvious that b^* is less than b_0, and increases with a. If the function $b^*(a)$ is known, it is easy to decide on the preferable renewal policy. If $b > b^*$ an age replacement policy is adopted, while if $b < b^*$ a block replacement policy is preferable. The decision rule is presented in Fig. 5.17. The function $b^*(a)$ is determined in the following way. The optimal ARP is computed for a given value of a, yielding the minimal average cost rate $k^*_{\mathrm{ARP}}(a)$. From relation (5.163) we compute the optimal period T^* of a block replacement policy which has the same minimal average cost rate as the ARP :

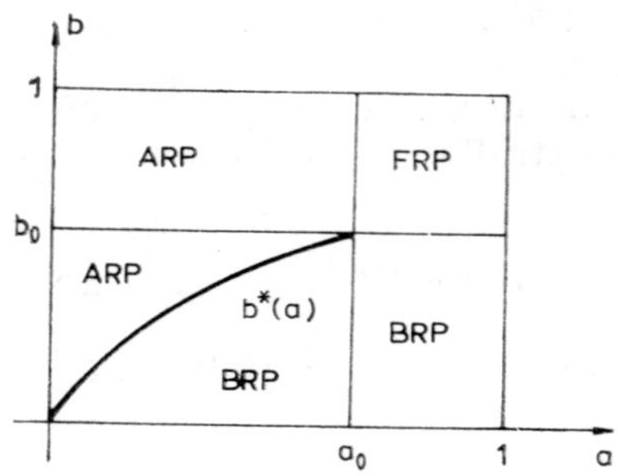

Fig. 5.17. — Selection of the optimal renewal policy.

$$k^*_{\mathrm{BRP}} = h(T^*) = k^*_{\mathrm{ARP}}$$
$$T^* = h^{-1}\,(k^*_{\mathrm{ARP}}). \tag{5.194}$$

Introducing the value T^* given by relation (5.194) into (5.162) we obtain the cost b^* for which ARP and BRP have the same minimal average cost rates :

$$b^* = T^*h(T^*) - H(T^*). \tag{5.195}$$

The cost b is compared to the cost b^* given by relation (5.195), and the decision is made. For the example considered throughout this Section, the existence of optimal age and block replacement policies require that $a_0 = 1/3$ and $b_0 = 2/9$. To compute the function $b^*(a)$, we equate expressions (5.165) and (5.191)

$$k^*_{\mathrm{BRP}} = \frac{2\lambda}{3}(1 - \mathrm{e}^{-y^*}) = k^*_{\mathrm{ARP}} = 2\lambda\,\xi(a) \tag{5.196}$$

where $y^* = 3\lambda T^*$ and $\xi(a) = (1 - a)\dfrac{a + \sqrt{a(4 - 3a)}}{2 - a + \sqrt{a(4 - 3a)}}$. Solving equation (5.196) for y^*, we get :

$$y^* = \ln\frac{1}{1 - 3\,\xi(a)}\cdot \tag{5.197}$$

The expression (5.197) for y^* is introduced into equation (5.162) for T^*, so that b^* can be computed as:

$$b^* = T^*h(T^*) - H(T^*) = \frac{2}{9} - \frac{2}{9}\, y^*\mathrm{e}^{-y^*} - \frac{2}{3}\, \mathrm{e}^{-y^*} =$$

$$= \frac{2}{3}\, \xi(a) + \frac{2}{9}[1 - 3\xi(a)] \ln[1 - 3\xi(a)]. \tag{5.198}$$

For $a = 0.25$ and $b = 0.2$, expression (5.198) yields $b^* = 1.1\%$, so the optimal age replacement policy is preferable.

The above discussion has shown that the renewal policies are efficient for positive aging systems (IFR). The models described have assumed that aging is a continuous process, that is, that the failure rate is a continuous increasing function. We shall now consider the case when the system failure rate has positive jumps at random instants, so that aging is a discrete stochastic process. This applies in redundant systems. Failure of some components does not imply system failure, but the system changes from its "good as new" state into a "wear out" state with a higher failure probability. Aging is then the discrete process of passing through the intermediate states, from the initial state to the failure state.

The two-unit parallel system is a good example, as failure of either unit makes the system pass into the "wear out" state, with one unit working and the other failed. This is an intermediate state, between the initial state (both units working) and the failed state (both units failed).

The best renewal policy for such systems is to eliminate the wear of the system as soon as it appears, that is, the failed redundant elements should be repaired before failure of the whole system actually occurs. This implies continuous monitoring of the system so that any component failure can be detected. When such monitoring is possible, we have a continuous replacement policy (CRP). According to this policy, the system is restored to its initial state either when it has failed or when it has reached a worn state.

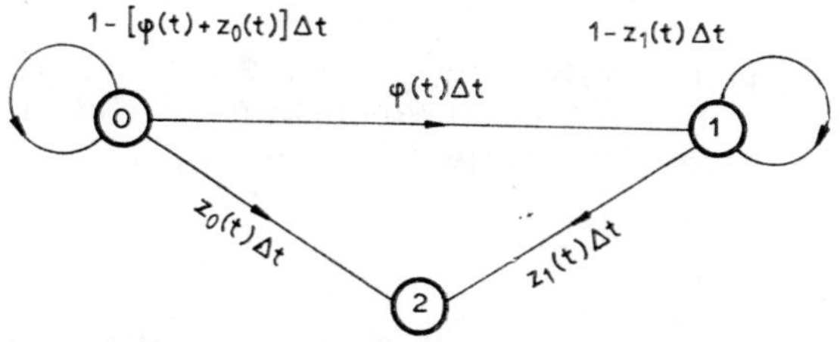

Fig. 5.18. — Markov chain model for system wear.

When continuous monitoring is not possible, the system is inspected periodically, with period T_E, preventive renewal being carried out if the system is found to be in a worn state upon inspection. This is an eventual replacement policy (ERP). Analysis of CRP and ERP requires a model for the process of wearing of the system. The model is the Markov chain, shown in Fig. 5.18. The system states are: the "good as new" state (0), the "wear out" state (1) and the failed state (2). $z_0(t)$ and $z_1(t)$ are the failure rates of the system in the "good as new" and "wear out" states, respectively. System wear is brought about by component failure in $(t, t + \Delta t)$, with probability $\varphi(t)\Delta t$. The equations of the states are :

$$
\begin{aligned}
P_0(t + \Delta t) &= P_0(t)[1 - (z_0(t) + \varphi(t))\, \Delta t] \\
P_1(t + \Delta t) &= P_0(t)\, \varphi(t)\, \Delta t + P_1(t)\, [1 - z_1(t)\Delta t].
\end{aligned}
\tag{5.199}
$$

Equations (5.199) are easily transformed into the differential equations :

$$
\begin{aligned}
\frac{\mathrm{d}P_0(t)}{\mathrm{d}t} &= - z_0(t) + \varphi(t))\, P_0(t) \\
\frac{\mathrm{d}P_1(t)}{\mathrm{d}t} &= \varphi(t)\, P_0(t) - z_1(t)\, P_1(t).
\end{aligned}
\tag{5.200}
$$

Solving equations (5.200) with initial conditions $P_0(0) = 1$ and $P_1(0) = 0$, we get :

$$
\begin{aligned}
P_0(t) &= \exp\left\{-\int_0^t [z_0(u) + \varphi(u)]\, \mathrm{d}u\right\}, \\
P_1(t) &= \exp\left[-\int_0^t z_1(u)\,\mathrm{d}u\right]\int_0^t \varphi(u)\exp\left\{-\int_0^u [z_0(v) + \varphi(v) - z_1(v)]\,\mathrm{d}v\right\}\mathrm{d}u.
\end{aligned}
\tag{5.201}
$$

The reliability function of the system is the sum of the probabilities associated with the working states, as the failed state is considered an absorbing state in the model. It is interesting to study the case when the transition rates are constant, so that :

$$
\varphi(t) = \varphi\,; \quad z_0(t) = \lambda_0\,; \quad z_1(t) = \lambda_1. \tag{5.202}
$$

The failure rate of the system has two levels, λ_0 and λ_1. The failure rate jumps from λ_0 to λ_1 with the constant transition rate φ. In this case, equation (5.201) of the states becomes :

$$
\begin{aligned}
P_0(t) &= \exp[-(\lambda_0 + \varphi)t] \\
P_1(t) &= \frac{\varphi}{\varphi + \lambda_0 - \lambda_1}\{\exp[-\lambda_1 t] - \exp[-(\lambda_0 + \varphi)t]\}.
\end{aligned}
\tag{5.203}
$$

The reliability function is then :

$$
R(t) = P_0(t) + P_1(t) = \frac{\varphi}{\varphi + \lambda_0 - \lambda_1}\mathrm{e}^{-\lambda_1 t} - \frac{\lambda_1 - \lambda_0}{\varphi + \lambda_0 - \lambda_1}\mathrm{e}^{-(\lambda_0 + \varphi)t}. \tag{5.204}
$$

From the reliability function (5.204), we can derive all the reliability measures useful in the study of CRP and ERP. Integrating expression (5.204) gives the mean time-to-failure :

$$m = \int_0^\infty R(t)\,\mathrm{d}t = \frac{\varphi + \lambda_1}{\lambda_1(\varphi + \lambda_0)}. \tag{5.205}$$

The probability density function of the time-to-failure is :

$$f(t) = -\frac{\mathrm{d}R(t)}{\mathrm{d}t} = \frac{(\varphi + \lambda_0)(\lambda_0 - \lambda_1)}{\varphi + \lambda_0 - \lambda_1}\,\mathrm{e}^{-(\lambda_0+\varphi)t} + \frac{\lambda_1\varphi}{\varphi + \lambda_0 - \lambda_1}\,\mathrm{e}^{-\lambda_1 t}. \tag{5.206}$$

The renewal density is :

$$h^*(s) = \frac{f^*(s)}{1 - f^*(s)} = \frac{s\lambda_0 + \lambda_1(\varphi + \lambda_0)}{s(s + \varphi + \lambda_1)}$$

$$h(t) = \frac{\lambda_1(\varphi + \lambda_0)}{\varphi + \lambda_1} - \frac{(\lambda_1 - \lambda_0)}{\varphi + \lambda_1}\,\mathrm{e}^{-(\varphi+\lambda_1)t}. \tag{5.207}$$

Integrating the renewal density (5.207), we get the renewal function

$$H(t) = \int_0^t h(t)\,\mathrm{d}t = \frac{\lambda_1(\varphi + \lambda_0)\,t}{\varphi + \lambda_1} + \frac{\varphi(\lambda_1 - \lambda_0)}{(\varphi + \lambda_1)^2}\,\mathrm{e}^{-(\varphi+\lambda_1)t} - \frac{\varphi(\lambda_1 - \lambda_0)}{(\varphi + \lambda_1)^2}. \tag{5.208}$$

The results are valid when no preventive renewal is carried out. The average maintenance cost rate is :

$$k_{\mathrm{FRP}} = \frac{1}{m} = \frac{\lambda_1(\varphi + \lambda_0)}{\varphi + \lambda_1}. \tag{5.209}$$

In the case of the continuous replacement policy with negligible renewal duration, the system is instantaneously changed from state (1) to state (0), and it may fail only when it is in the "good as new" state. The Markov chain model of the system under CRP is depicted in Fig. 5.19. A simple analysis yields the reliability function and the renewal density :

$$R(t) = \mathrm{e}^{-\lambda_0 t}; \quad h(t) = \lambda_0. \tag{5.210}$$

The renewal upon failure process is of the Poisson type, with density λ_0. The preventive renewal process is also of the Poisson type with density φ.

The average maintenance cost rate of a system under CRP is easily computed. Let c be the cost of a preventive renewal. As a failure occurs in $(t,t+\Delta t)$ with probability $\lambda_0 \Delta t$ and a preventive renewal is carried out in $(t,t+\Delta t)$ with probability $\varphi \Delta t$, the average maintenance cost rate is :

$$k_{\mathrm{CRP}} = \lambda_0 + c\varphi. \tag{5.211}$$

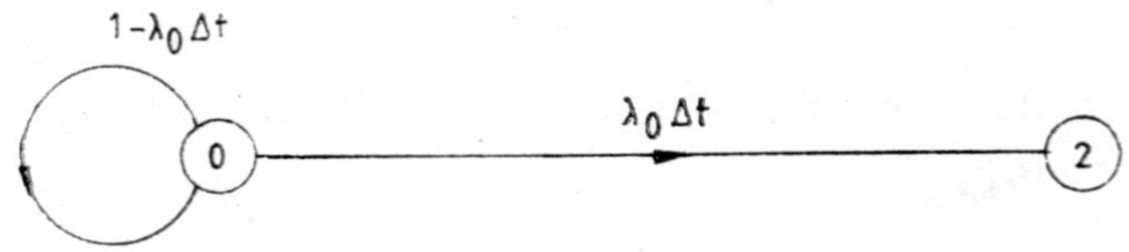

Fig. 5.19. — Markow chain model for a worn system under CRP.

Comparing the cost rates (5.209) and (5.211), we get the condition of efficiency of the continuous replacement policy :

$$\lambda_0 + c\varphi < \frac{\lambda_1(\varphi + \lambda_0)}{\lambda_1 + \varphi};$$
$$c < \frac{\lambda_1 - \lambda_0}{\lambda_1 + \varphi}. \tag{5.212}$$

Let us now consider the eventual replacement policy. The system is inspected periodically with period T_E, and is returned to its initial state if it is found to be in a worn state. As system failure is obvious, a repair is carried out whenever it is necessary. In the time interval between two successive inspections, the system is characterized by the reliability measures (5.204) — (5.208) derived under FRP. The mean cost of the renewals upon failure in a time interval T_E is then $H(T_E)$ and may be computed according to relation (5.208). Preventive renewal is carried out only when the system is in state (1) at T_E, with probability $P_1(T_E)$. This probability cannot be computed according to relations (5.203), because these were derived assuming an absorbing failed state, that is, without taking into account the renewals upon failure. This assumption was necessary in order to compute the reliability function.

The Markov chain model in the case of instantaneous system renewal upon failure is depicted in Fig. 5.20.

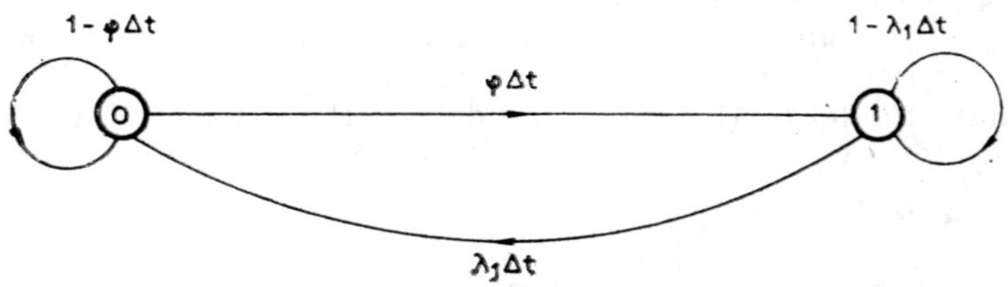

Fig. 5.20. — Markov chain model for a system under ERP.

Note that, because of the instantaneous renewals upon failure, the system may be in either state (0) or state (1), its failure restoring the system into state (0) irrespective of the state it was in before failure occurred.

The probability of the system being in state (1) is given by the equations :

$$\begin{aligned} P_1(t + \Delta t) &= P_0(t)\, \varphi \Delta t + P_1(t)\,(1 - \lambda_1 \Delta t) \\ P_0(t) + P_1(t) &= 1. \end{aligned} \tag{5.213}$$

From equations (5.213) we derive the differential equation :

$$\frac{\mathrm{d}P_1(t)}{\mathrm{d}t} = \varphi - (\lambda_1 + \varphi)\, P_1(t). \tag{5.214}$$

Integrating equation (5.214) with the initial condition $P_1(0) = 0$, we get :

$$P_1(t) = \frac{\varphi}{\lambda_1 + \varphi}\,[1 - \mathrm{e}^{-(\lambda_1+\varphi)t}]. \tag{5.215}$$

Let v be the cost of a preventive renewal under ERP. The average maintenance cost rate is :

$$k_{\mathrm{ERP}} = \frac{H(T_E) + vP_1(T_E)}{T_E}. \tag{5.216}$$

Introducing expressions (5.208) and (5.215) into (5.216) we get :

$$k_{\mathrm{ERP}} = \frac{\lambda_1(\varphi + \lambda_0)}{\varphi + \lambda_1} + \frac{\varphi[v\varphi + \lambda_0 - \lambda_1(1 - v)]}{T_E(\lambda_1 + \varphi)^2}[1 - \mathrm{e}^{-(\lambda_1+\varphi)T_E}]. \tag{5.217}$$

Taking the limits of k_{ERP} for $T_E \to 0$ and for $T_E \to \infty$ we get the average cost rates for CRP and FRP respectively :

$$\begin{aligned} \lim_{T_E \to 0} k_{\mathrm{ERP}}(T_E) &= v\varphi + \lambda_0 = k_{\mathrm{CRP}} \\ \lim_{T_E \to \infty} k_{\mathrm{ERP}}(T_E) &= \frac{\lambda_1(\varphi + \lambda_0)}{\varphi + \lambda_1} = k_{\mathrm{FRP}}. \end{aligned} \tag{5.218}$$

A period T_E between inspections may be derived such that k_{ERP} should be minimum. Setting the first derivative of k_{ERP} equal to zero we get the equation for T_E^* :

$$T_E^* h(T_E^*) - H(T_E^*) = v[P_1(T_E^*) - T_E P_1'(T_E^*)]. \tag{5.219}$$

As a matter of fact, when a redundant system with CFR components is maintained under a block or an age replacement policy, it is virtually an eventual replacement policy that is being carried out. Periodically, with period T^* or when the system reaches age x^*, the system is tested and restored to its initial state by replacing the failed redundant components.

Nothing is done to the working components, as they have not worn out, and nothing is done to the system, if all its components are working. Mere inspection assures that the system resumes its initial state.

Let us again take the example of the parallel system. Under BRP, the system is tested periodically with period T^*. If both units are working, the system is in state (0) and no renewal is performed. If a unit has failed, it is replaced by a new one and the system starts again in state (0). Similarly, under ARP the system is tested when it reaches age x^*, and the same operations are performed.

If the failure rates and the transition rate to the worn state are known, the period between inspections may be computed from relation (5.219), taking into account also (5.208) and (5.215). In the case of the parallel system, we have $\lambda_0 = 0$, $\lambda_1 = \lambda$ and $\varphi = 2\lambda$. Equation (5.219) becomes :

$$\frac{6 - 2v - 4w^*}{9} = \frac{2 - 6v}{9} \mathrm{e}^{w^*} \text{ where } w^* = 3\lambda\, T_E^*. \tag{5.220}$$

A solution of equation (5.220) exists for $v < 1/3$.

The average maintenance cost rates for the parallel system under FRP, CRP and ERP may now be obtained from expressions (5.209), (5.211) and (5.217) respectively :

$$\begin{aligned} k_{\text{FRP}} &= \frac{2\lambda}{3} \\ k_{\text{CRP}} &= 2\lambda c \\ k_{\text{ERP}} &= \frac{2\lambda}{3} + \frac{2(3v - 1)\,(1 - \mathrm{e}^{-3\lambda T_E})}{9T_E}. \end{aligned} \tag{5.221}$$

From expressions (5.221) we conclude that the CRP and ERP are preferable to the failure replacement policy when $c < 1/3$ and $v < 1/3$. All renewal policies may be compared when a unitary design criterion is used. When more design criteria are used, a decision must be based on the optimization theory with multiple goals. Table 5.2 presents the parameters of the renewal policies, as applied to the parallel system.

Recent developments of renewal policies are concerned with imperfect renewal [22, 23] and new cost models [24, 25, 26].

The renewal policies have been presented in this Section from a global point of view. Considerations of system structure lead to further development of some of these policies. The assumptions on which the renewal policies are based are simplistic, so it is to be expected that a theoretical optimal policy should be sub-optimal in practice. However, this does not diminish the value of mathematical models, as it would be too costly to base maintenance policies only on empirical observations.

Table 5.2

Renewal policies

Renewal policy	Domain	Parameter	Cost
FRP	—	—	$k_{FRP} = \frac{2\lambda}{3}$
BRP	$b < 2/9$	T^* $\left(1 - \frac{9b}{2}\right) e^{3\lambda T^*} = 1 + 3\lambda T^*$	$k^*_{BRP} = \frac{2\lambda}{3}(1 - e^{-3\lambda T^*})$
DRP	$b < 2/9$ $d \in (0, \infty)$	T^*, T_D^* same T^* as in BRP $T_D^* = \frac{1}{d}$	$k^*_{DRP} = k^*_{BRP}\left(1 - \frac{1}{2dT^*}\right)$
ARP	$a < 1/3$	x^* $x^* = \frac{1}{\lambda} \ln \frac{2(1-a)}{2 - 3a - \sqrt{a(4-3a)}}$	$k^*_{ARP} = 2\lambda(1-a) \frac{a + \sqrt{a(4-3a)}}{2 - a + \sqrt{a(4-3a)}}$
CRP	$c < 1/3$	—	$k_{CRP} = 2\lambda c$
ERP	$v < 1/3$	$\frac{6 - 2v - 12\lambda T_E^*}{9} = \frac{2 - 6v}{9} e^{3\lambda T_E^*}$	$k^*_{ERP} = \frac{2\lambda}{3} + \frac{2(3v-1)(1 - e^{-\lambda T_E^*})}{9T_E^*}$

References

1. TILLMAN, F.A., HWANG, C.L., KUO, W., System effectiveness models : an annotated bibliography, IEEE Trans. Reliab., R-29, 4(1980) 295—304.
2. COX, D.R., Renewal Theory, John Wiley, New York, 1962.
3. PLESSER, K.T., FIELD, T.O., Cost-optimized burn-in durations for repairable electronic systems, IEEE Trans. Reliab., R-26, 3(1977) 195—197.
4. SRINIVASAN, V.S., Tests for "New different than repaired", IEEE Trans. Reliab., R-27, 4(1978) 280—282.
5. MIHALACHE, A., On an Apparent Paradox in Renewal Theory (in Romanian). Buletinul IPB, 48(1986), 97—102.
6. KOLARIK, W.J., CASE, K.F., Hazard modification techniques for repair and multiple operating conditions, IEEE Trans. Reliab., R-29, 4(1980) 324—326.

7. SHANTHIKUMAR, J.G., A general software reliability model for performance prediction, Microelectron. Reliab., 21, 5(1981) 671—682.
8. GOEL, A.L., OKUMOTO, K., Time dependent error-detection rate model for software reliability and other performance measures, IEEE Trans. Reliab., R-28, 3(1979) 206—211.
9. LITTLEWOOD, B., VERRALL, J., A bayesian reliability model with a stochastically monotonic failure rate, IEEE Trans. Reliab., R-23, 2(1974) 108—115.
10. CĂTUNEANU, V.M., MIHALACHE, A., Improving the accuracy of the Littlewood-Verrall model, IEEE Trans. Reliab., R-34, 5(1985) 418—422.
11. LITTLEWOOD, B., How to measure software reliability and how not to., IEEE Trans. Reliab., R-28, 2(1979) 103—110.
12. NAKAGAWA, T., GOEL, A.L., A note on availability for a finite interval, IEEE Trans. Reliab., R-22, 5(1973) 271—272.
13. NAKAGAWA, T., OSAKI, S., Off-time distributions in an alternating renewal process with reliability applications, Microelectron, Reliab., 13(1974) 181—184.
14. MIHOC, GH., NICULESCU, S., Stochastical Renewal Processes (in Romanian), Ed. Academiei, Bucharest, 1982.
15. HEIKKILLĂ, M.A., Approximating the interval reliability by the use of instantaneous failure rates, IFAC, Helsinki, 1978, pp. 1629—1633.
16. HWANG, C.L., TILLMAN, F.A., WEI, W.K., LIE, C.H., Optimal scheduled-maintenance policy based on multiple-criteria decision-making, IEEE Trans. Reliab., R-28, 5(1979) 394—399.
17. BERG, M., EPSTEIN, B., Comparison of age, block and failure replacement policies, IEEE Trans. Reliab., R-27, 1 (1978) 25—27.
18. NAKAGAWA, T., YASUI, K., Approximate calculation of block replacement with Weibull failure times, IEEE Trans. Reliab., R-27, 4(1978) 268—271.
19. TANGO, T., A modified block replacement policy using less reliable items, IEEE Trans. Reliab., R-28, 5(1979), 400—401.
20. CHAN, P.K.W., DOWNS, T., Two criteria for preventive maintenance, IEEE Trans. Reliab., R-27, 4(1978) 272—273.
21. BERGMAN, B., KLEFSJÖ, B., A graphical method applicable to age-replacement problems, IEEE Trans. Reliab., R-31, 5(1982) 478—480.
22. NAKAGAWA, T., Imperfect preventive maintenance, IEEE Trans. Reliab., R-28, 5(1979) 402.
23. NAKAGAWA, T., Optimum policies when preventive maintenance is imperfect, IEEE Trans. Reliab., R-28, 4(1979) 331—332.
24. GHEORGHE, A.V., System Engineering (in Romanian), Ed. Academiei, Bucharest, 1979.
25. IOSIFESCU, M., GRIGORESCU, S., OPRIȘAN, GH., POPESCU, GH., Elements of Stochastic Modelling (in Romanian), Ed. tehnică, Bucharest, 1984.
26. KAIO, N., OSAKI, S., Some remarks on optimum inspection policies, IEEE Trans. Reliab., R-33, 4(1984) 277—283.
27. SHERWIN, D.J., Improved schedules by using data collected under preventive maintenance, IEEE Trans. Reliab., R-33, 4(1984) 315—320.
28. BEICHELT, F., A replacement policy based on limits for the repair cost rate, IEE Trans. Reliab., R-31, 4(1982) 401—402.
29. NAKAGAWA, T., A modified block replacement with two variables, IEEE Trans. Reliab., R-31, 4(1982) 398—400.
30. NAKAGAWA, T., Modified periodic replacement with minimal repair at failure, IEEE Trans. Reliab., R-30, 2(1981) 165—168.
31. BEICHELT, F., A generalized block-replacement policy, IEEE Trans. Reliab., R-30, 2(1981) 171—172.

Chapter 6

Structural reliability

6.1 Functional and logical models

In previous sections, reliability was modelled without regard to the system structure, by a global statistical approach. At the global level, the system is viewed as an abstract object determined by a set of variables which are its only connections to the outside world. The abstract system is analogous to the physical system if the condition of causality is fulfilled, which implies a classification of the variables into input (cause) and output (effect) variables. The input variables are components of the input vector $U = (u_1, u_2, \ldots, u_m)$ and the output variables are components of the output vector $Y = (y_1, y_2, \ldots, y_p)$. Both vectors are stochastic non-stationary processes. When the system structure is ignored, the global model is given by the functional relationship between the input and the output vectors:

$$Y = A(U). \tag{6.1}$$

Knowledge of the system structure indicates a set of internal variables which channel the influence of the input variables on the output variables. The internal variables are components of the state vector $\Gamma = (\gamma_1, \gamma_2, \ldots, \gamma_N)$. They are affected by the input variables and influence in turn the output variables. Consequently, the system is split into two subsystems such that the state variables are outputs of the first subsystem and inputs for the second subsystem (Fig. 6.1).

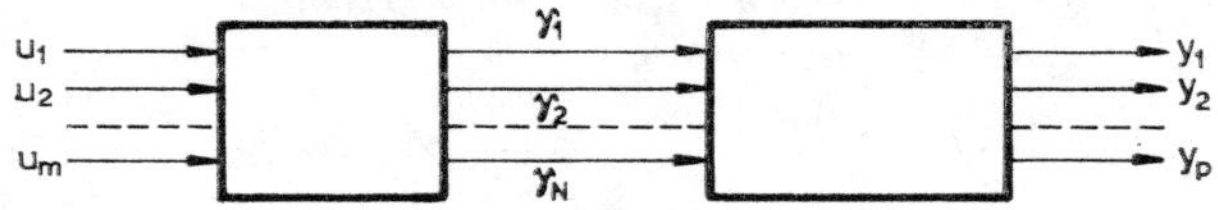

Fig. 6.1. — Input, output and state variables.

In quite general conditions [1,2], the system may be described by the state equations:

$$\begin{aligned} \dot{\Gamma} &= B(\Gamma, U); \\ Y &= C(\Gamma, U). \end{aligned} \tag{6.2}$$

There are many ways to choose the state variables. It is convenient to make a choice such that the subsystem which describes the state-output relation should be trivial. In this case, the state equations become :

$$\begin{aligned} \dot{\Gamma} &= B(\Gamma,\, U) \\ Y &= D(\Gamma). \end{aligned} \tag{6.3}$$

According to relations (6.3), the state vector completely determines the output vector. The effects of the input variables occur only via the intermediate state variables.

Reliability analysis of a real system is based on an adequate physical interpretation of the input, output and state variables. As shown in the first Chapter, it is natural to consider as output variables the system performances, and as input variables the stresses to which the system is subjected because of its interaction with other systems and with its environment. The choice of the state vector is based on the fact that the system performances depend directly on the component parameters which, in their turn, depend on the stresses. In a reliability analysis, the component parameters are then interpreted as state variables.

A complete reliability analysis should describe, mathematically, the influence of stresses on component parameters and the dependence of system performance on component parameters.

In many real situations the stresses may influence the system performance directly, but we shall neglect this so that relations (6.3) remain true.

The system performance, the component parameters and at least some of the stresses are stochastic processes, so that the structural reliability model is a multi-variable stochastical system.

The dependence of the component parameters on the stresses was described in Chapter 4. We shall deal in this Chapter with the dependence of the system performances on the component parameters. We shall develop reliability models so that reliability measures for the system may be derived from the reliability measures of its components. The structural reliability models are all based on the general relationship

$$Y = D(\Gamma). \tag{6.4}$$

The functional relationship (6.4) may be written more explicitly :

$$\begin{aligned} y_i &= f_i(\gamma_1^{(1)},\, \gamma_1^{(2)},\ldots,\, \gamma_1^{(l_1)},\, \gamma_2^{(1)},\, \gamma_2^{(2)},\ldots,\, \gamma_2^{(l_2)},\ldots,\, \gamma_n^{(1)},\, \gamma_o^{(2)},\ldots,\, \gamma_n^{(ln)}) \\ i &= 1,\, 2,\ldots,\, p. \end{aligned} \tag{6.5}$$

Relations (6.5) are a functional model of the system, as they relate the system performance to the component parameters. One notices that each component is described by l_j parameters $(j = 1,\, 2,\ldots,\, n)$ so that the total number of parameters is :

$$N = l_1 + l_2 + \ldots + l_n. \tag{6.6}$$

To define a reliability measure of the system, it is necessary to specify the admissible domain of the output vector in the p-dimensional space. We shall assume that it is possible to specify completely the failure criteria of the system, so that its satisfactory state may be expressed by the following inequalities imposed on its performance :

$$y_{i_L} \leqslant y_i \leqslant y_{i_U} \quad i = 1, 2, \ldots, p. \tag{6.7}$$

The interval (y_{i_L}, y_{i_U}) is a tolerance interval for performance i. It is, of course, possible that for some i, $y_{i_L} = 0$ or $y_{i_U} \to \infty$.

The reliability function of the system is then the probability that, during time interval $(0, t)$, all performances are within the respective tolerance intervals :

$$R_S = P\left[\bigcap_{i=1}^{p} (y_{i_L} \leqslant y_i \leqslant y_{i_U})\right]. \tag{6.8}$$

Starting from the tolerance intervals imposed on the system performances, the tolerence intervals associated with the component parameters may be computed by the methods described in [4] and [5]. They are defined by the inequalities :

$$\gamma_{j_L}^{(k)} \leqslant \gamma_j^{(k)} \leqslant \gamma_{j_U}^{(k)} ; \\ k = 1, 2, \ldots, l_j ; \quad j = 1, 2, \ldots, n. \tag{6.9}$$

The reliability function of a component j is the probability that, during time interval $(0, t)$ all its l_j parameters are within their respective tolerance intervals :

$$R_j = P\left[\bigcup_{k=1}^{l_j} (\gamma_{j_L}^{(k)} \leqslant \gamma_j^{(k)} \leqslant \gamma_{j_U}^{(k)})\right] \\ j = 1, 2, \ldots, n. \tag{6.10}$$

It is essential to note that the component reliability function (6.10) is defined with respect to extrinsic failure criteria which are imposed by the system performance and structure. It would be quite erroneous to adopt for R_j the values which may be found in a catalogue or in a standard, as those values are determined on the basis of intrinsic failure criteria which do not account for the role the component plays, and the place it occupies, in the structure of a specific system.

A first goal of structural reliability analysis is to establish a relationship between the system reliability function (6.8) and the component reliability functions (6.10).

The analysis must be preceded by a realistic evaluation of the individual reliabilities R_j, taking into account the actual failure criteria of the components imposed by the system structure.

To solve this problem one must know, at each instant, the probability density functions associated with the component parameters $\{\gamma_j^{(k)};\ k = 1, 2, \ldots, l_j;\ \ j = 1, 2, \ldots, n\}$. From these pdfs, the component reliability functions are computed according to relations (6.10). On the other hand, the pdf associated with $\gamma_j^{(k)}$ may in principle be derived according to the structural relation (6.5), yielding the probability density functions associated with the system performaces. Then, from relation (6.8), the system reliability function is obtained. Repeating this operation for different time instants, the functional dependence between R_s and $\{R_j;\ j = 1, 2, \ldots, n\}$ is established.

The main theoretical difficulty of this procedure is the blending of the component parameter pdfs in order to get the pdfs of the system performances. To avoid this difficulty, Monte Carlo simulation should be used in order to generate possible specific values of $\gamma_j^{(k)}$ according to their respective pdfs. For each set of specific values of $\gamma_j^{(k)}$, specific values of the system performances y_i may be computed according to relation (6.5). If the number of simulations is large enough, the pdfs associated with the system performances may be derived with sufficient accuracy [6,7]. Such computer-aided analysis has the advantage of being closely connected with the system design, but it is time consuming becuse of the need to repeat the analysis for different time instants. Moreover, this repeated analysis may be carried out only by assuming complete knowledge of the stochastic processes $\{\gamma_j^{(k)};\ k = 1, 2, \ldots, l_j;\ j = 1, 2, \ldots, n\}$, that is, the pdfs of $\gamma_j^{(k)}$ must be known for any time instant. Such knowledge requires a large number of measurements, performed repeatedly over a long time span [7]. That is why direct structural analysis based on the functional model (6.5) is usually avoided.

A simplified structural model may be derived from the tolerance intervals (6.7) and (6.9) associated with the system performance and the component parameters, respectively, and from the structural relations (6.5). Instead of p output variables, a binary variable $S(t)$ is defined so that, for each time instant t:

$$\begin{aligned} &S(t) = 1 \ \text{ if } \forall i = 1, 2, \ldots, p : y_{i_L} \leqslant y_i \leqslant y_{i_U} \\ &S(t) = 0 \ \text{ if } \exists i : y_i < y_{i_L} \text{ or } y_i > y_{i_U}. \end{aligned} \tag{6.11}$$

According to definition (6.11), the output vector is replaced by a binary variable which takes the value "1" if all performances are within tolerances and the value "0" if at least one of the performances is outside its tolerance interval.

We shall restrict the analysis to prediction of the first system failure, without taking into account the renewal possibilities of the system. In this case, the system reliability function in $(0, t)$ is equal to the probability that the system is in a working state at time instant t, that is:

$$R_s(t) = P[S(t) = 1]. \tag{6.12}$$

Similarly, instead of the l_j-dimensional state vector associated with component j, a binary variable x_j is defined so that it is equal to "1" if all the component parameters are within tolerances and to "0" in the opposite case.

$$x_j(t) = 1 \text{ if } \forall k = 1, 2, \ldots, l_j : \gamma_{jL}^{(k)} \leqslant \gamma_j^{(k)} \leqslant \gamma_{jU}^{(k)}$$

$$x_j(t) = 0 \text{ if } \exists k : \gamma_j^{(k)} < \gamma_{jL}^{(k)} \text{ or } \gamma_j^{(k)} > \gamma_{jU}^{(k)}. \tag{6.13}$$

The reliability function of component j is then :

$$R_j(t) = P[x_j(t) = 1]. \tag{6.14}$$

The system state at instant t is given by the binary vector $(x_1, x_2, \ldots, x_n)$. The structural relations (6.5) may be used to classify the system states into "working" and "failed" ones. Consequently, the state-output relationship may be expressed by the boolean structure function

$$S = \varphi(x_1, x_2, \ldots, x_n). \tag{6.15}$$

The structure function (6.15), together with definitions (6.11) and (6.13), define the logical model. Relation (6.15) indicates the logical combinations of component states which imply satisfactory behaviour of the system. Let us denote two system states by X_i and X_j. These states may be ordered so that $X_i \geqslant X_j$ if at least one component variable takes a "1" value in state i instead of a "0" value in state j. The system is coherent if, for any two states $X_i \geqslant X_j$, we have

$$S_i = \varphi(X_i) \geqslant S_j = \varphi(X_j). \tag{6.16}$$

A coherent system behaves better as more of its components are in working condition. It is easy to see that the structure function of a coherent system should not contain operators such as NOT or EXCLUSIVE OR.

The goal of the structural analysis is to derive, from the logical model (6.15), a functional relationship between the system reliability function (6.12) and the component reliability functions (6.14) :

$$R_S = \psi(R_1, R_2, \ldots, R_n). \tag{6.17}$$

The logical model simplifies the reliability analysis for two reasons. Firstly, it assumes complete knowledge of the individual reliability functions with respect to the actual failure criteria imposed on the components by the system performance and structure. Thus, reliability analysis based on the logical model does not require a preliminary assessment of the component reliability functions from the pdfs of the component parameters. In such cases, when the actual component reliabilities are not known, the numerical result of the reliability analysis is erroneous, but it is still useful in comparing different structures which employ the same components.

Secondly, the logical model describes the system structure by the boolean function (6.11), which is simpler than the equations (6.5) of the functional model. This simplification results from the fact that relations (6.5) describe the system from all points of view, while the function (6.11) describes only its stuctural reliability by indicating the combinations of component states which engender satisfactory behaviour of the system.

We shall apply the functional and the logical model in reliabiltiy analysis of a control system [8]. A control system with negative feedback is depicted in Fig. 6.2.

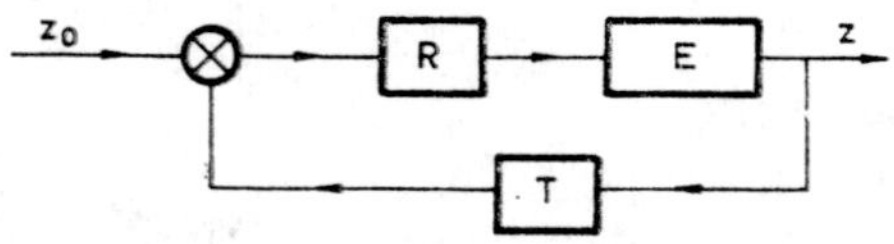

Fig. 6.2. — Feedback loop controi system: E — technological equipment; T — transducer; R — controller.

The purpose of the control system is to keep the ouput signal $z(t)$ constant. The system performance is defined as the relative variation of the output signal. Thus the performance vector is one-dimensional, with component

$$y = \frac{\Delta z}{z}. \tag{6.18}$$

The system components are grouped as in Fig. 6.3.

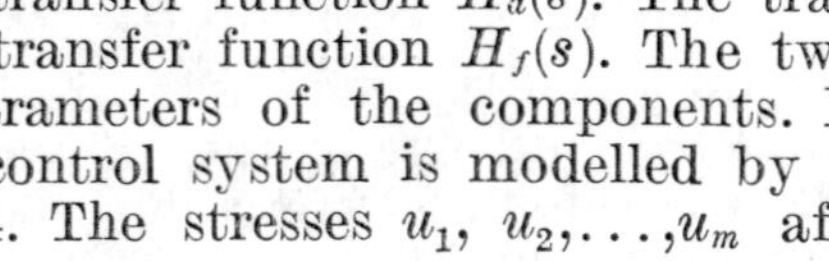

Fig. 6.3. — Control system structure.

The controller and the technological equipment form the direct path element, with transfer function $H_d(s)$. The transducer is the feedback element, with transfer function $H_f(s)$. The two transfer functions are the relevant parameters of the components. From the reliability point of view, the control system is modelled by the abstract system depicted in Fig. 6.4. The stresses u_1, $u_2, \ldots, u_m$ affect the parameters

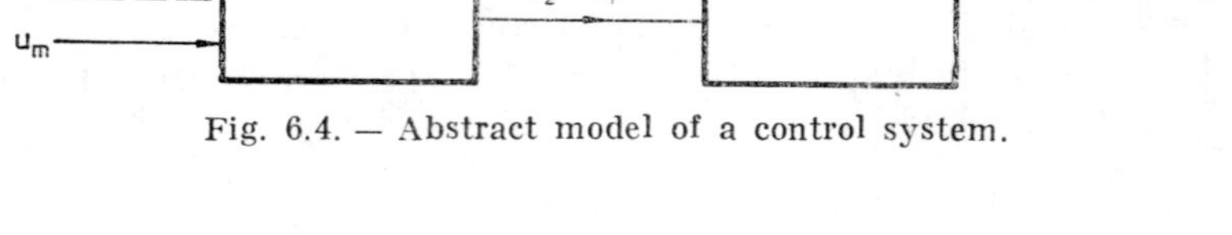

Fig. 6.4. — Abstract model of a control system.

H_d and H_f of the components and these, in their turn, influence the system performance.

In order to derive a structural reliability model, a state-output equation of the type (6.4) is needed. Such an equation may be obtained from the transfer function of the system

$$\frac{z}{z_0} = \frac{H_d}{1 + H_d H_f}. \tag{6.19}$$

The reference z_0 is assumed constant.

By differentiation of the logarithm of expression (6.19), we get

$$\frac{dz}{z} = \frac{1}{1 + H_d H_f} \frac{dH_d}{H_d} - \frac{H_d H_f}{1 + H_d H_f} \frac{dH_f}{H_f}. \tag{6.20}$$

Relation (6.20) may be written with finite differences :

$$\frac{\Delta z}{z} = \frac{1}{1 + H_d H_f} \frac{\Delta H_d}{H_d} - \frac{H_d H_f}{1 + H_d H_f} \frac{\Delta H_f}{H_f}. \tag{6.21}$$

Equation (6.21) defines the functional model of the control system, as it relates the system performance to the component parameters.

The system reliability function is the probability that the relative variation of the output signal does not exceed some prescribed value ε :

$$R_S = P(|y| \leqslant \varepsilon). \tag{6.22}$$

The tolerances associated with the component parameters may be computed by using a classical method [4,5], for instance the worst case design method. The condition imposed on the system performance may be written :

$$|y| = \frac{|\Delta z|}{z} \leqslant \frac{1}{1 + H_d H_f} \frac{|\Delta H_d|}{H_d} + \frac{H_d H_f}{1 + H_d H_f} \frac{|\Delta H_f|}{H_f} \leqslant \varepsilon. \tag{6.23}$$

Let $\varepsilon_d + \varepsilon_f = \varepsilon$. From inequality (6.23), we get

$$\begin{aligned} \frac{|\Delta H_d|}{H_d} &\leqslant \varepsilon_d(1 + H_d H_f) \\ \frac{|\Delta H_f|}{H_f} &\leqslant \varepsilon_f \frac{1 + H_d H_f}{H_d H_f} \approx \varepsilon_f. \end{aligned} \tag{6.24}$$

Inequalities (6.24) define the tolerance intervals of the component parameters. The component reliability functions are

$$\begin{aligned} R_d &= P\left[\frac{|\Delta H_d|}{H} \leqslant \varepsilon_d(1 + H_d H_f)\right] \\ R_f &= P\left[\frac{|\Delta H_f|}{H_f} \leqslant \varepsilon_f\right]. \end{aligned} \tag{6.25}$$

We shall use the functional model to derive the relationship between the system reliability function (6.22) and the component reliability functions (6.25). For this purpose, the following steps must be carried out :

1. The pdfs $f_{H_d}(H_d)$ and $f_{H_f}(H_f)$ associated with the parameters H_d and H_f are established;
2. The pdfs derived in step (1) are introduced into relations (6.25), in order to compute the component reliability functions;
3. The pdf associated with the output signal z is computed, using relation (6.19) and the simulated values for H_d and H_f;
4. The system reliability function is computed according to relation (6.22);
5. Steps (1) to (4) are repeated for different time instants.

It should be noted that an analytical solution is not possible, because of the difficulty in deriving theoretically the pdf of the output signal (step 3). The Monte Carlo simulation of specific values for H_d and H_f is expensive in computer time and memory, so it may be concluded that reliability analysis based on the functional model is impractical.

To derive the logical model of the control system, we define the binary variable S which takes the value "1" if $|y| \leqslant \varepsilon$ and the value "0" in the opposite case. We define also the binary variables x_d and x_f which take respectively, the values "1" or "0" if the inequalities (6.24) are satisfied, or not. One notices that the system performance lies within the tolerance interval (6.22) if and only if the inequalities (6.24) are satisfied. Consequently, both components should be in working order to keep the system in a working state. The structure function is

$$S = x_d x_f. \tag{6.26}$$

The structure function (6.26) is much simpler than equation (6.21), which defines the functional model. If the components are independent, the system reliability function may be written as :

$$R_S = P(S = 1) = P(x_d = 1)\ P(x_j = 1) = R_d R_f. \tag{6.27}$$

The simplicity of a reliability analysis based on the logical model arises from the fact that, firstly, the model describes only the reliability aspects of the system structure and, secondly, the actual component reliabilities are assumed known.

Suppose now that the component reliability functions are computed with respect to some intrinsic failure criteria, say :

$$R_d = P\left(\frac{\Delta H_d}{H_d} \leqslant \varepsilon_d\right); \quad R_f = P\left(\frac{\Delta H_f}{H_f} \leqslant \varepsilon_f\right). \tag{6.28}$$

Multiplying expressions (6.28), we obtain the system reliability with respect to an artificial failure criterion, resulting from the combination of the intrinsic failure criteria of the components. The reliability analysis will then ignore the actual requirements imposed on the system by relation (6.22). The logical model describes the system as a collection of independent components, ignoring the restrictions imposed upon them by the system structure. Comparing relations (6.28) and (6.25) for the same numerical values of ε_d and ε_f, we notice that the failure criterion of the transducer remains unchanged by its insertion into the system structure, while the actual failure criterion of the controller is less severe than the intrinsic one, due to the presence of the feedback loop. The modifications of the intrinsic failure criteria are due to the interactions determined by the structure of the system.

It is thus important in a reliability analysis to take into account adequate failure criteria for the components. The example of the control system has shown that a reliability analysis based upon the logical model is simple and adequate, provided that the system structure is taken into account when establishing the data concerning the components.

6.2 System reliability analysis based upon logical models

In this Section, we shall present the main methods used in the quantitative reliability analysis of a system described by the logical model. The goal of the analysis is to derive, from the structure function (6.15), a relationship of the type (6.17) between system and component reliability functions.

First, we shall associate with the logical function (6.15) a signal flow graph. To each component there corresponds at least one link in the graph. The links are interrupted when the corresponding components fail. The failed state of the system is symbolized in the graph (diagram) by the fact that the input signal does not reach the output node. If a one-to-one correspondence between components and links is required, the graph should be directed. If not, a non-directed graph can represent any coherent structure.

Let us consider the simple structure functions which use only the AND operator:

$$S = x_1 \cap x_2 \cap \ldots \cap x_n. \tag{6.29}$$

The system fails when any component fails.

The associated signal flow graph is shown in Fig. 6.5.

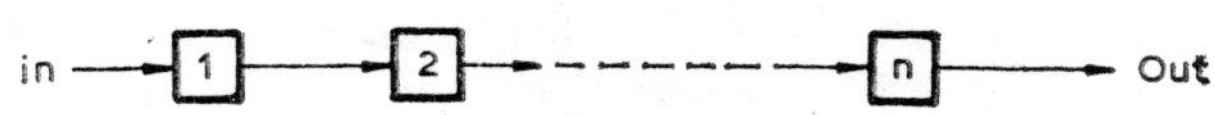

Fig. 6.5. — Series system.

Because of the aspect of the graph, such systems as described by the structure function (6.29) are called series systems, irrespective of their functional structure. The control system analysed in Section 6.1 is a series system from the reliability point of view, though it is a feedback system from the functional point of view.

Assuming that component failures are statistically independent, the system reliability function may be written:

$$R_S = P(S = 1) = \prod_{i=1}^{n} P(x_i = 1) = \prod_{i=1}^{n} R_i. \tag{6.30}$$

Relation (6.30) is a specific case of relation (6.15) for series systems. From relation (6.30), it may be concluded that a series system with CFR components is also CFR. We have, indeed:

$$R_S = \prod_{i=1}^{n} \exp[-\lambda_i t] = \exp\left[-t \sum_{i=1}^{n} \lambda_i\right] = \exp\ [-\lambda_s t]. \tag{6.31}$$

The constant failure rate and the MTTF of the series system are

$$\lambda_S = \sum_{i=1}^{n} \lambda_i; \quad m_S = \frac{1}{\sum_{i=1}^{n} \lambda_i}. \tag{6.32}$$

It may be seen that, if one component is much less reliable than the others, that is $\lambda_j \gg \lambda_i$; $\forall_{i \neq j}$, this component determines the system reliability ($\lambda_S \simeq \lambda_j$). This "principle of the weakest link" shows that it is not sensible to build up a series system from components which differ strongly in their reliabilities.

Let us now consider a structure function which uses only the OR operator:

$$S = x_1 \cup x_2 \cup \ldots \cup x_n. \tag{6.33}$$

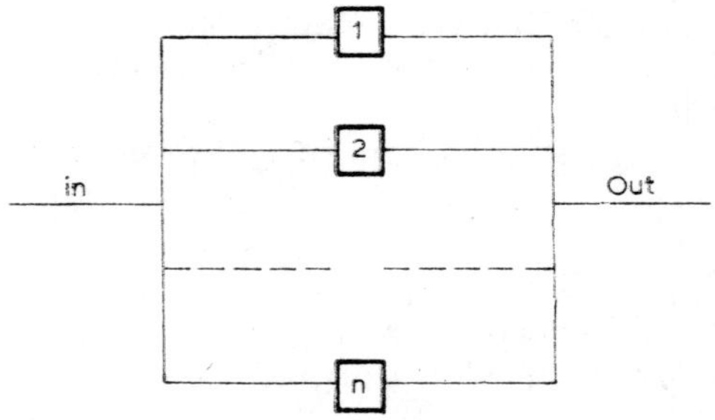

Fig. 6.6. — Parallel system.

The system fails when all its components fail. The graph associated with the structure function (6.33) is depicted in Fig. 6.6. Because of its aspect, such systems as described by the structure function (6.33) are called parallel systems, irrespective of their functional structure.

A quantitative analysis is straightforward if the negation of expression (6.33) is considered:

$$\bar{S} = \bar{x}_1 \cap \bar{x}_2 \cap \ldots \cap \bar{x}_n. \tag{6.34}$$

For statistically independent components, the system failure probability is the product of the component failure probabilities

$$F_S = P(\bar{S} = 1) = \prod_{i=1}^{n} P(\bar{x}_i = 1) = \prod_{i=1}^{n} F_i. \tag{6.35}$$

The parallel system reliability function is then :

$$R_S = 1 - F_S = 1 - \prod_{i=1}^{n} (1 - R_i). \tag{6.36}$$

Analysis of expression (6.36) shows that a parallel system with CFR components is IFR when the components are identical and IFRA when the components are not identical [9, 10, 11]. More generally, a parallel system with components of the IFR, IFRA or NBU type not necessarily identical is also of the IFR, IFRA or NBU type respectively [12]. The positive aging property of the parallel system with memoryless components is explained by the fact that failure of some components increases the failure probability of the system.

Once the elementary structures of series and paralell types have been analysed, it is straightforward to approach any system which is a combination of series and parallel structures, provided there is a one-to-one correspondence between components and links. However, this is not generally the case. The structure function of any coherent system contains AND and OR operators, such that it may be written in the s-o-p (sum of products) form

$$S = \bigcup_{i=1}^{m} (x_{i_1} \cap x_{i_2} \cap \ldots \cap x_{i_{ji}}). \tag{6.37}$$

Each term of the structure function is a path set, that is, a set of components which, when in a sound state, assure the satisfactory behaviour of the system. If the boolean function (6.37) is minimized, each term is a minimal path set. Any coherent system has a structure function of the form (6.37), that is a logical sum of minimal path sets. If the path sets have no common elements, the system may be analysed using relation (6.30) for each path set and relation (6.36) for the parallel arrangement of path sets. When different path sets have common elements, this approach is no longer appropriate, as the path sets which have common elements are not independent, so relation (6.36) does not hold any more.

The methods we shall subsequently describe are able to deal with cases when the path sets have one or more common elements. We shall restrict the analysis to coherent systems but, as will be pointed out, some of these methods are also adequate for systems which are not coherent.

Statistical independence of the components is an essential assumption. The modelling of statistical dependence of components is complicated and shows, more often than not, that the result obtained under the assumption of independence is a lower bound on the true system reliability function [13, 14, 15].

A first method, which is particularly suited to manual computations, is based on the total probability formula. According to this method, the system structure is simplified to a series-parallel form, with no common elements, by making hypotheses on the state of some of the components. Obviously, these hypotheses should refer to the common elements in different path sets. Let x_j be a binary variable which appears in several path sets. When x_j takes the value "1", component j is in working state and the structure function becomes

$$S/(x_j = 1) = \varphi(x_1, x_2, \ldots, x_{j-1}, 1, x_{j+1}, \ldots, x_n). \tag{6.38}$$

When x_j takes the value "0", component j has presumably failed and the structure function is

$$S/(x_j = 0) = (x_1, x_2 \ldots x_{j-1}, 0, x_{j+1}, \ldots x_n). \tag{6.39}$$

If the path sets of the simplified structures (6.38) and (6.39) do not contain any common elements their reliability functions may be readily computed. According to the total probability formula, the system reliability function is :

$$\begin{aligned} R_S &= P(S = 1) = P(S = 1/x_j = 1)P(x_j = 1) + \\ &+ P(S=1/x_j= 0) \cdot P(x_j = 0) = R(S/j)R_j + R(S/\bar{j})(1 - R_j). \end{aligned} \tag{6.40}$$

If the path sets of the simplified structures (6.38) and (6.39) still contain common elements, additional hypotheses are made concerning another element, say x_k. The conditional reliabilities are then :

$$\begin{aligned} R(S/j) &= R(S/j \cap k)R_k + R(S/j \cap \bar{k})(1 - R_k) \\ R(S/\bar{j}) &= R(S/\bar{j} \cap k)\, R_k + R(S/\bar{j} \cap \bar{k})(1 - R_k). \end{aligned} \tag{6.41}$$

It should be noted that the system structure may be simplified by formulating hypotheses on the state of a subsystem or on the state of a set of components. In the latter case, the system reliability function will have 2^m terms, where m is the number of elements in the set.

It is convenient to express, by the total probability method, the statistical importance a component has in ensuring the satisfactory behaviour of the system. The statistical importance of component j is defined as the derivative of the system reliability function with respect to the reliability function of component j :

$$I_j = \frac{\partial R_S}{\partial R_j} = R(S/j) - R(S/\bar{j}). \tag{6.42}$$

We shall consider, as an example, the communication system shown in Fig. 6.7, together with its logical model.

Each information source (a node in the functional model) is connected to the others by unreliable links. Let R denote the reliability function of a link. The system reliability function is the probability of information transmission between any two nodes. Because of the structural symmetry of the system, the logical model is the same for all node

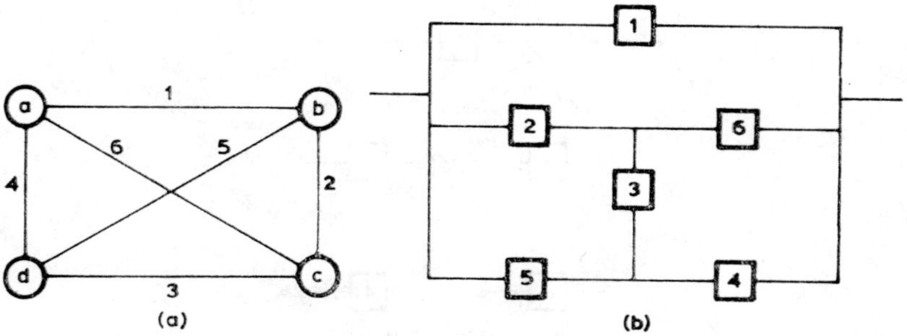

Fig. 6.7. — Communication system and its logical model.

pairs (a and b in Fig. 6.7). Moreover, the logical pattern coincides with the functional model as the actual and conventional failure modes are the same.

A logical analysis of the system yields the structure function written in the s-o-p form :

$$S = x_1 \cup x_2 x_6 \cup x_4 x_5 \cup x_2 x_3 x_4 \cup x_3 x_5 x_6 \tag{6.43}$$

where $x_i x_j \triangleq x_i \cap x_j$.

The graph associated with expression (6.43) is not a combination of series and parallel structures (Fig. 6.7). However, it could be transformed into a series-parallel one if more symbolic edges were used to represent the same actual link. In both cases, the direct method of computation cannot be used. To simplify the system structure, a hypothesis concerning the state of any of the components 2 to 6 may be made, as all the variables x_2 to x_6 are common to different path sets. Considering component 3, the structure function given that component 3 is working is :

$$S/(x_3 = 1) = x_1 \cup x_2\, x_6 \cup x_4 x_5 \cup x_2 x_4 \cup x_5 x_6 =$$

$$= x_1 \cup [(x_2 \cup x_5)(x_4 \cup x_6)]. \tag{6.44}$$

The graph associated with the simplified structure function (6.44) is obtained from the general diagram (graph) by short-circuiting edge 3, which means that the symbolic signal is transmitted with certainty through the corresponding link (Fig. 6.8).

The system is of the series-parallel type, so that its reliability function may be computed by repeatedly applying relations (6.30) and (6.36). We have:

$$R(S/3) = P(S = 1/x_3 = 1) = 1 - (1 - R)\{1 - [1 - (1 - R)^2]^2\} =$$
$$= - R^5 + 5R^4 - 8R^3 + 4R^2 + R. \tag{6.45}$$

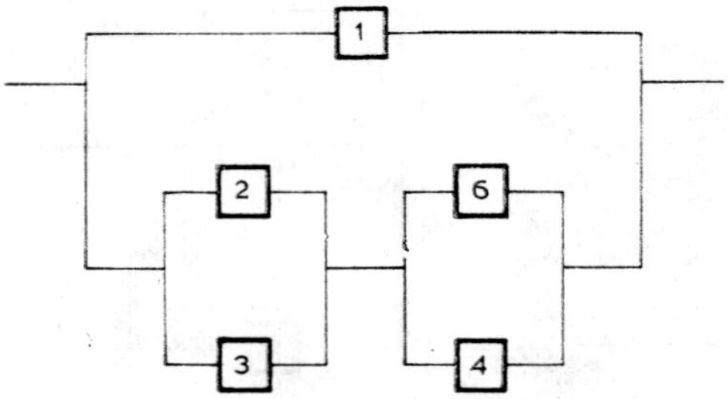

Fig. 6.8. — Graph associated with structure function (6.44).

If we now assume that link 3 has failed, the structure function becomes:

$$S/(x_3 = 0) = x_1 \cup x_2x_6 \cup x_4x_5. \tag{6.46}$$

Interrupting edge 3 in the general diagram, we get the graph associated with expression (6.46), which is of the series-parallel type (Fig. 6.9). The system reliability function, given that link 3 has failed, is:

$$R(S/\overline{3}) = P(S = 1/x_3 = 0) = 1 - (1 - R)(1 - R^2)^2 =$$
$$= R^5 - R^4 - 2R^3 + 2R^2 + R. \tag{6.47}$$

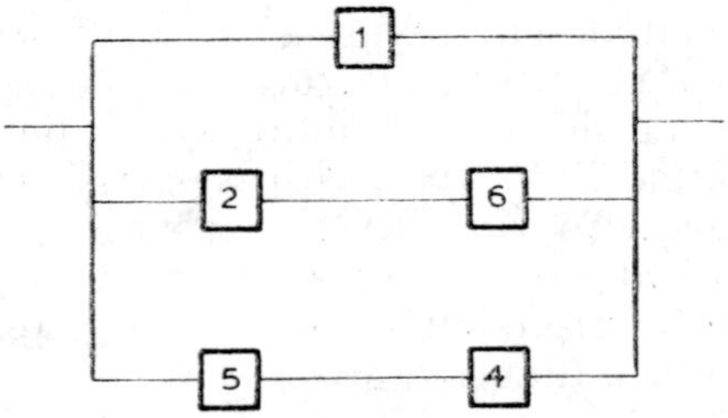

Fig. 6.9. — Graph associated with structure function (6.46).

We now apply the total probability formula in order to get the unconditioned system reliability function:

$$R_S = R(S/3)R_3 + R(S/\overline{3})(1 - R_3) =$$
$$= - 2R^6 + 7R^5 - 7R^4 + 2R^2 + R. \tag{6.48}$$

If we now assume that the links are identical CFR elements with failure rate λ, the mean system time-to-failure may readily be computed as:

$$m_S = \int_0^\infty R_S(t)\,\mathrm{d}t = \int_0^\infty (-2\mathrm{e}^{-6\lambda t} + 7\mathrm{e}^{-5\lambda t} - 7\mathrm{e}^{-4\lambda t} + 2\mathrm{e}^{-2\lambda t} +$$

$$+ \mathrm{e}^{-\lambda t})\,\mathrm{d}t = -\frac{2}{6\lambda} + \frac{7}{5\lambda} - \frac{7}{4\lambda} + \frac{2}{2\lambda} + \frac{1}{\lambda} = \frac{79}{60\lambda}. \tag{6.49}$$

We shall now compute the statistical importance of link 3. Introducing expressions (6.45) and (6.47) into relation (6.42) we get:

$$\frac{\partial R_S}{\partial R_3} = R(S/3) - R(S/\overline{3}) = -2R^5 + 6R^4 - 6R^3 + 2R^2 =$$

$$= 2R^2(1 - R)^3. \tag{6.50}$$

The statistical importance of element 3 is a function of time by the intermediary of R. It has a maximum value at $R = 2/5$ (Fig. 6.10).

The total probability method may be more conveniently applied when the logical diagram is used instead of the structure function. That is why often the logical diagram alone is derived, without specifying explicitly the structure function. This fact suggested the application to reliability of some methods of analysis used in electrical network theory, such as the delta-star conversion, which may reduce any structure to a series-parallel one. This method has conceptual difficulties concerning the independence of the elements of the resulting star connections. A controversy

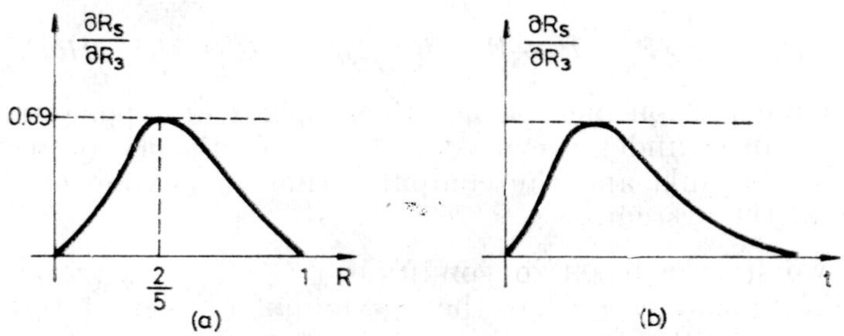

Fig. 6.10. — Statistical importance of link 3.

arose about the exactness of the transformations, due to imprecise definitions for the equivalence of the two sub-networks. A transformation which is exact, considering certain aspects of the equivalence, may turn out to be only approximate when some additional features of the equivalence are considered [16, 17, 18, 19].

In order to analyse the reliability of systems with greater complexity, systematic methods are needed which can easily be programmed on a computer. One such is the inclusion-exclusion method, based on the classical rule of computing the probabilities of a reunion of events [20].

Let the system be described, by the function (6.37), as a parallel structure of m path sets with common variables. Let P_1 be the sum of the probabilities of the m paths, P_2 the sum of the joint probabilities of the m paths taken two at a time, P_3 the sum of the joint probabilities of the paths taken three at a time, etc. According to Kolmogorov's axiom, extended for a reunion of m events, the system reliability function is :

$$R_S = P_1 - P_2 + P_3 - \ldots + (-1)^{m-1} P_m. \tag{6.51}$$

If there are m parallel paths with no overlapping components, expression (6.51) has the maximum possible number of terms, $2^m - 1$. However, if there are overlapping components (as is usually the case), the actual number of terms turns out to be a fraction of the maximum number because of cancellation. Because an inclusion alternates with an exclusion, the number of terms is odd and there is always exactly one more term with a plus sign than with a minus sign. If the equation has $2K + 1$ terms, then K terms have minus signs and $K + 1$ have plus signs.

Consider, as an example, the structure function (6.43) of the communication system. We have :

$$\begin{aligned} P_1 &= R + 2R^2 + 2R^3 ; \\ P_2 &= 2R^3 + 7R^4 + R^5 ; \\ P_3 &= 9R^5 + R^6 ; \\ P_4 &= 4R^6 + R^5 ; \\ P_5 &= R^6 ; \\ R_S &= P_1 - P_2 + P_3 - P_4 + P_5 = R + 2R^2 - 7R^4 + 7R^5 - 2R^6. \end{aligned} \tag{6.52}$$

The inclusion-exclusion method has the property of exponentially increasing time because, under the most extreme conditions, both the size of the resultant formula and the computer time are exponential functions of the size of the system.

The worst case has two conditions

(a) Expression (6.51) has the maximum number of terms, that is $2^m - 1$;

(b) Each term is the maximum possible size; if there are n components in the system, each term would have n variables.

These conditions would not occur in a real problem. Condition (b) is unlikely, because the algorithm is such that some terms are smaller than others. Similarly, if every term has the same components and all of them would have the same value, there would be extensive cancellation and condition (a) would not hold. If condition (a) were true, the system would be strictly parallel with no overlapping elements between paths;

this could be solved more easily by series and parallel reduction, rather than by overloading the computer with a general algorithm.

Another general method consists of expanding the structure function in a reunion of disjoint terms, each term being a logical product of binary variables. The sum of products (s—o—p) expression (6.37) becomes a sum of disjoint products (s—d—p) and the system reliability function is readily obtained, as the probability of a reunion of disjoint events is simply the sum of the probabilities of the terms.

Several methods have been developed in order to obtain an s—d—p form [22, 23, 24, 25]. The method we describe is one of the most efficient for both simple and complex systems.

Consider a structure function with two minimal path sets S_1 and S_2. It may be shown by a truth table that :

$$S = S_1 \cup S_2 = S_1 \cup \bar{S}_1 S_2. \tag{6.53}$$

Thus, a reunion of two terms may be expanded in a reunion of two disjoint terms. Generalizing for m path sets, we get :

$$S = \bigcup_{i=1}^{m} S_i = S_1 \cup \bar{S}_1 S_2 \cup \bar{S}_1 \bar{S}_2 S_3 \cup \ldots \cup \bar{S}_1 \bar{S}_2 \ldots \bar{S}_{m-1} S_m. \tag{6.54}$$

Relation (6.54) may be proved by induction. For $m = 2$, (6.54) reduces to (6.53), which is evidently true. Assume that relation (6.54) is true for $m - 1$. We can then write :

$$\begin{aligned} S &= \left(\bigcup_{i=1}^{m-1} S_i\right) \cup S_m = \left(\bigcup_{i=1}^{m-1} S_i\right) \cup \left(\overline{\bigcup_{i=1}^{m-1} S_i}\right) S_m = \\ &= (S_1 \cup \bar{S}_1 S_2 \cup \bar{S}_1 \bar{S}_2 S_3 \cup \ldots \cup \bar{S}_1 \bar{S}_2 \ldots \bar{S}_{m-2} S_{m-1}) \cup \\ &\cup (\bar{S}_1 \bar{S}_2 \ldots \bar{S}_{m-1} S_m) \end{aligned} \tag{6.55}$$

which proves relation (6.54).

If the path sets have common elements, one or more variables may appear repeatedly in the same term, so that the condition of independence between the variables in a logical product will no longer hold. Developing the logical products $\bar{S}_i$ according to De Morgan's rules does not always solve this difficulty as compatible terms would eventually arise. The following example clears up this point. Let $S_1 = xy$, $S_2 = xz$ and $S_3 = w$. According to relation (6.54), we may write :

$$S = S_1 \cup S_2 \cup S_3 = S_1 \cup \bar{S}_1 S_2 \cup \bar{S}_1 \bar{S}_2 S_3 = xy \cup \overline{xy}\, xz \cup \overline{xy}\, \overline{xz}\, w. \tag{6.56}$$

The last two terms of expression (6.56) contain the variable x twice, and these repetitions must be eliminated. Applying De Morgan's rules, we get :

$$S = xy \cup x\bar{y}z \cup \bar{x}w \cup \bar{x}\bar{z}w \cup \bar{x}\bar{y}w \cup \bar{y}\bar{z}w. \tag{6.57}$$

The last four terms in expression (6.57) are not disjoint.

To eliminate repetitions of a variable in a logical product we must use the identity

$$\overline{AB} = \bar{A} \cup \bar{B} = \bar{A} \cup A\bar{B} \tag{6.58}$$

which can be readily proved by De Morgan's rules and by expression (6.53). Developing the third term of expression (6.56) according to identity (6.58), we obtain a reunion of disjoint terms :

$$\overline{xy}\;\overline{xz}\;w = (\bar{x} \cup x\bar{y})(\bar{x} \cup x\bar{z})w = (\bar{x} \cup x\bar{y}\bar{z})w. \tag{6.59}$$

The other possible cases of variable duplication are $\overline{xy}\;xz$ and $\bar{x}\;\overline{xy}$, which may be treated by De Morgan's rules. Finally, we get the following rules for deleting a repeated variable :

$$\begin{aligned} \overline{yx} \cdot \overline{xz} &= \bar{x} \cup x\bar{y}\bar{z} \\ \overline{xy}\;xz &= x\bar{y}z \\ \bar{x}\;\overline{xy} &= \bar{x}. \end{aligned} \tag{6.60}$$

Once the structure function is of the s—d—p form with no repeating variables in any term, the reliability function is readily obtained by replacing each variable with its corresponding reliability function.

The procedure described above is efficient if, in the initial form of the structure function, the minimal sets are ordered according to their lengths, so that the first term is the one with the least number of factors. Moreover, the path sets with most common elements should be adjacent.

Consider, for instance, the structure function (6.43). After ordering the path sets, the structure function becomes :

$$S = S_1 \cup S_2 \cup S_3 \cup S_4 \cup S_5 \tag{6.61}$$

where $S_1 = x_1$; $S_2 = x_2x_6$; $S_3 = x_4x_5$; $S_4 = x_2x_3x_4$; $S_5 = x_3x_5x_6$.

The s—d—p form of the structure function (6.61) is

$$\begin{aligned} S &= S_1 \cup \bar{S}_1S_2 \cup \bar{S}_1\bar{S}_2S_3 \cup \bar{S}_1\bar{S}_2\bar{S}_3S_4 \cup \bar{S}_1\bar{S}_2\bar{S}_3\bar{S}_4S_5 = \\ &= S_1 \cup \bar{S}_1(S_2 \cup \bar{S}_2(S_3 \cup \bar{S}_3(S_4 \cup \bar{S}_4S_5))). \end{aligned} \tag{6.62}$$

A concise notation for expression (6.62) is :

$$S = \underset{3}{\underline{\qquad}}\; S_4 \cup \bar{S}_4S_5 \tag{6.63}$$

where $\underset{3}{\underline{\qquad}}$ signifies all operations in relation (6.62) up to $\bar{S}_3$ inclusively.

Introducing the expressions for S_4 and S_5 into relation (6.63), we get:

$$S = \underset{3}{\underline{\qquad}}\; x_2x_3x_4 \cup \overline{x_2x_3x_4}\; x_3x_5x_6. \tag{6.64}$$

The duplicate variable x_3 from the second term of expression (6.64) is deleted according to the rules (6.60), so that

$$S = \underset{3}{\text{———}}\, x_2x_3x_4 \cup \overline{x_2x_4}\, x_3x_5x_6. \tag{6.65}$$

Proceeding similarly with path set S_3, we get :

$$S = \underset{2}{\text{———}}\, x_4x_5 \cup \overline{x_4x_5}\,(x_2x_3x_4 \cup \overline{x_2x_4}x_3x_5x_6) =$$

$$= \underset{2}{\text{———}}\, x_4x_5 \cup x_2x_3x_4\,\overline{x_4x_5} \cup \overline{x_2x_4}\,\overline{x_4x_5}\,x_3x_5x_6 =$$

$$= \underset{2}{\text{———}}\, x_4x_5 \cup x_2x_3x_4\bar{x}_5 \cup x_3\bar{x}_4x_5x_6. \tag{6.66}$$

The next step is to introduce path set S_2 :

$$S = \underset{1}{\text{———}}\, x_2x_6 \cup \overline{x_2x_6}(x_4x_5 \cup x_2x_3x_4\bar{x}_5 \cup x_3\bar{x}_4x_5x_6) =$$

$$= \underset{1}{\text{———}}\, x_2x_6 \cup \overline{x_2x_6}\,x_4x_5 \cup x_2\,\overline{x_2x_6}x_3x_4\bar{x}_5 \cup$$

$$\cup\, x_3\bar{x}_4x_5x_6\,\overline{x_2x_6} = \underset{1}{\text{———}}\, x_2x_6 \cup \overline{x_2x_6}x_4x_5 \cup x_2x_3x_4\bar{x}_5\bar{x}_6 \cup$$

$$\cup\, \bar{x}_2x_3\bar{x}_4x_5x_6. \tag{6.67}$$

Finally, introducing path set S_1, we get the structure function expanded in the s$-$d$-$p form :

$$S = x_1 \cup \bar{x}_1(x_2x_6 \cup \overline{x_2x_6}x_4x_5 \cup x_2x_3x_4\bar{x}_5\bar{x}_6 \cup \bar{x}_2x_3\bar{x}_4x_5x_6) =$$

$$= x_1 \cup \bar{x}_1x_2x_6 \cup \bar{x}_1\overline{x_2x_6}x_4x_5 \cup \bar{x}_1x_2x_3x_4\bar{x}_5\bar{x}_6 \cup \bar{x}_1\bar{x}_2x_3\bar{x}_4x_5x_6. \tag{6.68}$$

Expression (6.68) is a reunion of disjoint terms, each term being an intersection of independent random variables. The system reliability function may be determined by inspection, replacing S by $R_S = P(S = 1)$ and x_i by $R_i = P(x_i = 1)$. We get :

$$R_S = R_1 + (1 - R_1)R_2R_6 + (1 - R_1)(1 - R_2R_6)R_4R_5 +$$

$$+ (1-R_1)R_2R_3R_4(1-R_5)(1-R_6) +$$

$$+ (1 - R_1)(1-R_2)R_3(1 - R_4)R_5R_6. \tag{6.69}$$

It should be noted that, unlike the inclusion-exclusion method, the s$-$d$-$p method yields the reliability function in an entirely additive form. There are no exclusions, and every term has a plus sign. For equal individual reliability functions, expression (6.69) reduces to

$$R_S = R + (1 - R)R^2 + (1 - R)(1 - R^2)R^2 + 2(1 - R)^3R^3 =$$

$$= -2R^6 + 7R^5 - 7R^4 + 2R^2 + R \qquad (6.70)$$

which coincides with the results obtained by the total probability and the inclusion-exclusion methods.

A useful feature of the s—d—p method for certain special cases is that if the paths are ordered so that every path differs from its immediate predecessor by exactly one variable, then the number of terms in the reliability expression is equal to the number of paths, and the formula is at most a polynomial function of the number of variables [20].

An algebraic method of reliability analysis [27] is based on the transformation of the logical structure function into an algebraic function, according to the equivalences between logical and algebraic operators presented in Table 6.1. These equivalences may readily be proved by truth tables.

Table 6.1

Logical-algebraic equivalences

Name	Logical function	Algebraic function
AND	$x \cap y$	xy
OR	$x \cup y$	$x + y - xy$
NOT	$\bar{x}$	$1 - x$
EXCLUSIVE OR	$x \oplus y$	$x + y - 2xy$

Let $S = \varphi(x_1, x_2, \ldots, x_n)$ be the logical structure function and $S = \psi(x_1, x_2, \ldots, x_n)$ its equivalent algebraic function. As S is a binary variable, its mean is equal to the probability of S being 1, that is, the system reliability function. Similarly, the mean of x_i is the reliability function of component i. As the algebraic function structure is a linear combination of products of independent variables, the system reliability may readily be obtained by taking the mean of $\psi(x_1, x_2, \ldots, x_n)$:

$$R_S = E_S = \psi(E_{x_1}, E_{x_2}, \ldots E_{x_n}) = \psi(R_1, R_2, \ldots R_n). \qquad (6.71)$$

Once the algebraic function structure is obtained, the reliability function is determined by inspection. It is important to note that this method does not require the structure function to be of s—o—p form. Moreover, as the logical operators NOT and EOR have equivalent algebraic operators, this method can also treat not-coherent systems.

We shall derive the algebraic equivalent function for the logical structure function (6.43). It is easier to consider the complement of expression (6.43) :

$$\bar{S} = \overline{x_1 \cup x_2x_6 \cup x_4x_5 \cup x_2x_3x_4 \cup x_3x_5x_6} =$$
$$= \bar{x}_1 \, \overline{x_2x_6} \, \overline{x_4x_5} \, \overline{x_2x_3x_4} \, \overline{x_3x_5x_6} . \tag{6.72}$$

Using Table 6.1, we get the algebraic function :

$$S = 1-(1-x_1)(1-x_2x_6)(1-x_4x_5)(1-x_2x_3x_4)(1-x_3x_5x_6). \tag{6.73}$$

The factors in expression (6.73) have common variables, so that they are not independent. Before taking the mean of expression (6.73), it is therefore necessary to develop it in an s—o—p form. After some computation, taking into account that $x_i^2 = x_i$, we get :

$$S = 1 - (1 - x_1)(1 - x_4x_5 - x_2x_6 - x_2x_3x_4 - x_3x_5x_6 +$$
$$+ x_2x_3x_4x_5 + x_3x_4x_5x_6 + x_2x_4x_5x_6 + x_2x_3x_4x_6 + x_2x_3x_5x_6 -$$
$$- 2x_2x_3x_4x_5x_6). \tag{6.74}$$

We may now take the mean of expression (6.74), as there are no more products of dependent variables. For identical components, the system reliability function is :

$$R_S = 1 - (1 - R)(1 - 2R^2 - 2R^3 + 5R^4 - 2R^5) =$$
$$= - 2R^6 + 7R^5 - 7R^4 + 2R^2 + R \tag{6.75}$$

which coincides with the results obtained by the other methods.

A method developed recently is that of topological reliability analysis [28, 29, 31, 32]. The idea of this method is that there is one-to-one correspondence between certain subgraphs, called the *p*-acyclic subgraphs of the logical model, and the non-cancelling terms of the system reliability expression. A *p*-graph is a subgraph of the logical model with every edge of the subgraph on a path, denoting success, from the source vertex to the terminal vertex. A *p*-acyclic graph is a *p*-graph with no cycles.

If we can identify the *p*-acyclic subgraphs, we can also save processing time in deriving the reliability expression because there will be no duplications or cancellations.

The system graph is decomposed by systematically stripping away edges and sequences of edges to find the *p*-acyclic subgraphs. The system reliability expression is generated simultaneously. Instead of a sum of monomial terms, as in the inclusion-exclusion method, the expression is in a nested and factored form that is equivalent, factor to term, to the polynomial obtained by inclusion-exclusion. Each factor also has the same sign as in the corresponding inclusion-exclusion term.

The method was devised for directed graphs but it can also be applied to non-directed graphs, replacing every edge which goes both ways by two one-way edges in opposite directions.

Another method, called the dynamic method [34] uses the concept of trajectory, which is a succession of states through which the system passes from the initial state to the failed state. A minimal cut set of reliability graph is a trajectory, but the concept of a trajectory extends beyond the concept of a cut set. The cut set is a static concept which defines a failed state, while the trajectory is a dynamic concept which defines the evolution of the system, by a succession of component failures, to a failed state. Therefore, unlike the cut sets, the trajectories are necessarily disjoint even if they have common elements, as the system has to follow only one trajectory during its evolution to a failed state. For instance, a parallel system has one minimal cut set (all elements failed) and $n!$ trajectories, because there are $n!$ possible sequences of failure of n components.

Let w_i be a trajectory and $P(w_i)$ the probability that the system follows it. As the trajectories are disjoint, we have

$$\sum_{i=1}^{N} P(w_i) = 1 \tag{6.76}$$

where N is the number of trajectories. Let $F(t/w_i)$ be the cumulative distribution function of the time required for the system to follow trajectory w_i. According to the total probability, the system failure probability in $(0, t)$ is

$$F(t) = \sum_{i=1}^{N} F(t/w_i)\, P(w_i). \tag{6.77}$$

Expression (6.77) may be easily computed for systems with CFR components.

Let $\{\lambda_k\}$ be the failure rates of the components which are good in a specified system state. Consider an adjacent state, where component k has failed. The transition rate between these two states is λ_k/Λ_k, where Λ_k is the sum of the failure rates of the good components in the first state. The probability of a trajectory w_i is then

$$P(w_i) = \prod_{k \in w_i} \frac{\lambda_k}{\Lambda_k}. \tag{6.78}$$

The state duration is exponentially distributed with probability Λ_k. The trajectory duration is the sum of state durations, so that its distribution is the convolution of the exponential distributions associated with the state durations. Consequently, the Laplace transform of $F(t/w_i)$ is:

$$\mathscr{L}F(t/w_i) = \frac{1}{s} \prod_{k \in w_i} \frac{\Lambda_k}{s + \Lambda_k}. \tag{6.79}$$

Taking into account expressions (6.78) and (6.79), the Laplace transform of the failure probability (6.77) becomes :

$$\mathscr{L}F(t) = \frac{1}{s}\sum_{i=1}^{N}\prod_{k\in w_i}\frac{\lambda_k}{s+\Lambda_k}. \tag{6.80}$$

When the components are identical, relations (6.78), (6.79) and (6.80) become extremely simple.

The trajectories may be grouped in M classes according to their lengths. Each class contains m_k trajectories of length k. The probability of a trajectory of length k is :

$$P(w_k)= \frac{1}{n(n-1)\ldots(n-k+1)} = \frac{1}{A_n^k}. \tag{6.81}$$

The cdf of the duration of trajectory w_k is

$$\mathscr{L}F(t/w_k)= \frac{1}{s}\,\frac{n\lambda}{s+n\lambda}\cdot\frac{(n-1)\lambda}{s+(n-1)\lambda}\cdots\frac{(n-k+1)}{s+(n-k+1)}. \tag{6.82}$$

Finally, the Laplace transform of the system failure probability is :

$$\mathscr{L}F(t) = \sum_{k=1}^{M}\frac{m_k\lambda^k}{s(s+n\lambda)[s+(n-1)\lambda]\ldots[s+(n-k+1)\lambda]}. \tag{6.83}$$

Relation (6.83) shows that the system failure probability may be computed directly if the number of trajectories of equal length is known.

When the components are not identical, it is necessary to specify all trajectories, which is rather difficult as they are numerous even for moderately complex systems.

We shall apply this method to derive the failure probability of the communication system used as an example throughout this Chapter. Inspecting the reliability graph from Fig. 6.7, we derive the number of trajectories :

$$m_1 = 0;\ m_2 = 0;\ m_3 = 2\cdot 3! = 12;\ m_4 = 2\cdot 4! = 48;$$
$$m_5 = 5\cdot 4! = 120;\ m_6 = 5! = 120. \tag{6.84}$$

Introducing the values (6.84) into relation (6.83) and applying the inverse Laplace transform, we obtain the expression of the failure probability, which is consistent with the results obtained by the other methods.

When the system structure is very complex, analytical derivation of the system reliability function becomes cumbersome. A numerical evolution of the system reliability function may be performed in such cases by a simulated experiment. The method consists of generating possible states of the components according to their reliability functions, and evaluating the system state of each combination of individual states. The ratio between the resulting number of system good states and the total number of simulations is a point estimate of the system reliability function.

The accuracy of this estimate increases with the number of simulations.

The point estimate of the system reliability function is

$$\hat{R}_S = \frac{\sum_{k=1}^{N} S_k}{N} \tag{6.85}$$

where N is the number of simulations and S_k is the value of the structure function for simulation k. Let R be the true reliability function and $W = \sum_{k=1}^{N} S_k$ the number of successes in a sequence of N trials. W is binomially distributed, with mean RN and standard deviation $\sqrt{RN(1 - R)}$. The accuracy of the estimate (6.85) may be expressed by the skewness, that is, the ratio between the standard deviation and the mean of W.

$$\frac{\sigma_W}{E_W} = \frac{\sqrt{RN(1-R)}}{RN} = \sqrt{\frac{1-R}{RN}}. \tag{6.86}$$

Relation (6.86) states that, for a specified number of simulations, estimations of higher reliability are more accurate. Moreover, the accuracy, and hence the number of simulations, does not depend on system complexity.

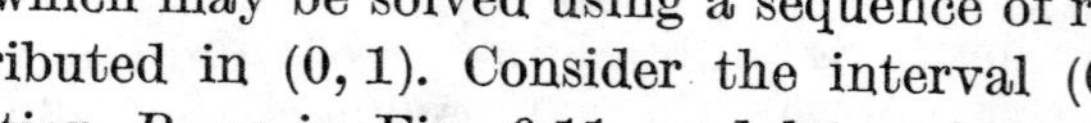

Fig. 6.11. — Simulation of x_i.

Simulation of component states means generating values of binary variable x_i according to its distribution law. This is a typical Monte Carlo problem which may be solved using a sequence of random numbers, uniformly distributed in (0, 1). Consider the interval (0, 1) divided by reliability function R_i as in Fig. 6.11, and let z_k be a random number uniformly distributed in (0, 1). It is obvious that

$$\begin{aligned} P(z_k < R_i) &= R_i = P(x_i = 1) \\ P(z_k \geqslant R_i) &= 1 - R_i = P(x_i = 0). \end{aligned} \tag{6.87}$$

According to relation (6.87), a "1" value is assigned to the variable x_i if $z_k < R_i$ and a "0" value in the opposite case. A specific value x_{i_k} is then obtained. For $k = 1, 2, \ldots, N$, we shall get N realizations of x_i. Similarly, we find specific values for the other individual variables.

Table 6.2

Monte Carlo simulation of the states of a communication system

	1	2	3	4	5	6	S
$(z/x)_1$	0.86/0	0.51/1	0.59/1	0.07/1	0.95/0	0.66/1	1
$(z/x)_2$	0.15/1	0.56/1	0.64/1	0.34/1	0.56/1	0.55/1	1
$(z/x)_3$	0.81/0	0.23/1	0.32/1	0.94/0	0.37/1	0.75/0	0
$(z/x)_4$	0.78/0	0.02/1	0.69/1	0.18/1	0.60/1	0.33/1	1
$(z/x)_5$	0.93/0	0.42/1	0.50/1	0.29/1	0.92/0	0.24/1	1
$(z/x)_6$	0.88/0	0.95/0	0.55/1	0.37/1	0.58/1	0.91/0	1
$(z/x)_7$	0.64/1	0.11/1	0.88/0	0.67/1	0.41/1	0.68/1	1
$(z/x)_8$	0.64/1	0.21/1	0.63/1	0.85/0	0.18/1	1.13/1	1
$(z/x)_9$	0.89/0	0.67/1	0.33/1	0.18/1	0.17/1	0.26/1	1
$(z/x)_{10}$	0.64/1	0.53/1	0.80/0	0.70/1	0.06/1	0.07/1	1

We shall apply the method to the communication system with the structure function (6.43). We assume that the six components have the same reliability function $R = 0.75$. Simulation of the state vector $(x_1, x_2, \ldots, x_6)$ proceeds according to Table 6.2, which contains ten specific values for $(x_1, x_2, \ldots, x_6)$ computed on the basis of the random numbers given in [35]. The values S_k which correspond to the system states are computed, for each realization of $(x_1, x_2, \ldots, x_6)$ according to the structure function (6.43).

The point estimate of the system reliability function is

$$\hat{R}_S = \frac{\sum_{k=1}^{10} S_k}{10} = \frac{9}{10} = 0.9. \tag{6.88}$$

The true reliability function may be computed by incorporating in the exact reliability expression (6.75) the value $R = 0.75$. We get $R_s = 0.9664$, which implies a relative error of 6.64%. The estimation accuracy for 10 simulations is, from relation (6.86), equal to 5.89%, which is close to the observed error.

When the number of simulations needs to be large, more sophisticated methods are necessary to reduce computer time and memory requirements [36, 37]. Monte Carlo simulation was also used in the context of the functional model, where component parameters were simulated according to their pdfs. In the logical model the procedure is simpler, as binary variables only are simulated.

A major limitation of methods based on the logical model is that they cannot treat systems that have renewal possibilities. A general model applicable to systems with renewal will be described in the next Section.

6.3 The Markov chain model

Let us consider a coherent system with structure function $\varphi(x_1, x_2, \ldots, x_n)$, Associating with each system state $(x_1, x_2, \ldots, x_n)$ a real number $s(t)$. we can define a discrete stochastic process with some remarkable properties.

The set of specific values of the stochastic process is finite and numerable. Indeed, the system has a finite number of components, so that the number of combinations of their states is finite, and the system states are distinct as they differ from each other at least by the state of a component.

Any modification of the stochastic process $s(t)$ occurs by the failure or completion of the repair of a component in a time interval $(t, t + \Delta t)$. It is natural to assume that the probability of two or more components changing their states in $(t, t + \Delta t)$ is negligible. Thus, only adjacent states are accessible from one another, and the process is ordinary.

Let s_0 be the system state at time t_0. The probability that the system state at $t > t_0$ is s depends on s_0, because state s is reached from s_0 by a finite number of transitions. However, this probability does not depend on the way the system has reached state s_0, before time t_0. The trajectory of the system from t_0 to t is independent of its trajectory up to t_0. This property is called the absence of after-effect.

The above assumptions describe a Markow chain, that is, a discrete ordinary stochastic process, without after-effect.

The Markov chain model is quite general, as it can describe system reliability either at the global or at the structural level, eventually taking into account the possibilities of renewal of the system.

We shall consider first the Markov chain model for a system, taking no account of either its structure or its renewal resources. The system has two states, good and failed, labelled "0" and "1" respectively. They can be represented as nodes of the topological graph in Fig. 6.12. The branches correspond to the transition probabilities. The transition from state "0" to state "1" occurs in $(t, t + \Delta t)$ with probability $z(t)\Delta t$, where $z(t)$ is the system failure rate. The reliability analysis consists of computing the state probabilities and, hence, the system reliability measures. The probability that the system is in state "0" at $t + \Delta t$ is :

$$P_0(t + \Delta t) = P_0(t)(1 - z(t)\Delta t). \tag{6.89}$$

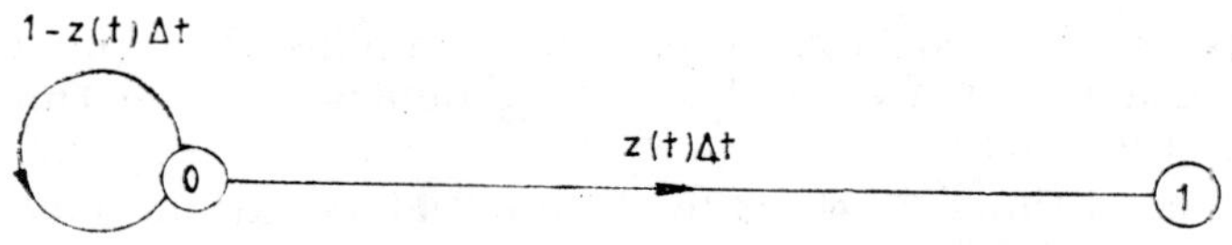

Fig. 6.12. — Global Markov chain for non-repairable systems.

Equation (6.89) is easily transformed into the differential equation

$$\frac{dP_0(t)}{dt} = - z(t)\, P_0(t). \tag{6.90}$$

Solving equation (6.90) with the initial condition $P_0(0) = 1$, we get :

$$P_0(t) = \exp\left[- \int_0^t z(u)\, du \right]. \tag{6.91}$$

For non-repairable systems, the probability $P_0(t)$ is precisely the reliability function, that is, a non-repairable system is up at t if and only if it has been up during $(0, t)$. This result was already known from the global analysis in Chapter 2.

Assume now that the system is repairable with repair rate $\rho(t)$. The system may recover to state "0" with probability $\rho(t)\Delta t$. The Markov chain of a repairable system is shown in Fig. 6.13.

The state probabilities at $t + \Delta t$ are

$$\begin{aligned} P_0(t + \Delta t) &= P_0(t)(1 - z(t)\Delta t) + P_1(t)\rho(t)\Delta t \\ P_0(t) + P_1(t) &= 1. \end{aligned} \tag{6.92}$$

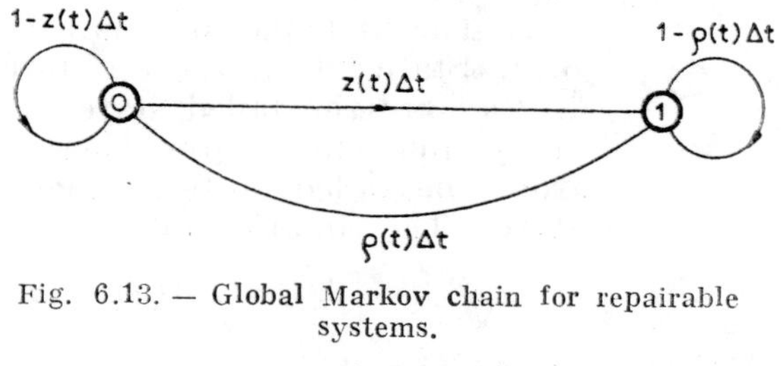

Fig. 6.13. — Global Markov chain for repairable systems.

Equations (6.92) convert into the differential form

$$\frac{dP_0(t)}{dt} = - [z(t) + \rho(t)]P_0(t) + \rho(t). \tag{6.93}$$

A solution of equation (6.93) is easy to find when both the time-to-failure and the time-to-repair are exponentially distributed, that is, the failure and repair rates are constant :

$$z(t) = \lambda$$
$$\rho(t) = \rho. \tag{6.94}$$

The assumption of a constant repair rate implies that the probability of completing a repair around t does not depend on the time spent in carrying out the repair.

Under conditions (6.94), the differential equation (6.93) is easily solved, yielding for initial condition $P_0(0) = 1$:

$$P_0(t) = \frac{\rho}{\lambda + \rho} + \frac{\lambda}{\lambda + \rho} e^{-(\lambda+\rho)t}. \tag{6.95}$$

$P_0(t)$ is the probability that the system is up at t, that is, $P_0(t)$ is the system availability. This reliability measure could be derived by the general methods of renewal theory (Chapter 5), so that the Markov chain model is superfluous in the case of global level analysis.

We shall now derive the Markov chain model for an elementary series structure. Consider a series system of n components, each having failure rate λ_i and repair rate ρ_i. The mean time-to-failure and the mean time-to-repair of component i are

$$m_i = \frac{1}{\lambda_i}; \quad \mu_i = \frac{1}{\rho_i}. \tag{6.96}$$

Two Markov chains have to be considered. In the first, the system renewal is overlooked, as at this step of the analysis the system reliability function and the mean time-to-failure have to be derived. The second Markov chain takes account of the recovery transitions and is used to derive the renewal function, the availability, the interval reliability and the mean residual life.

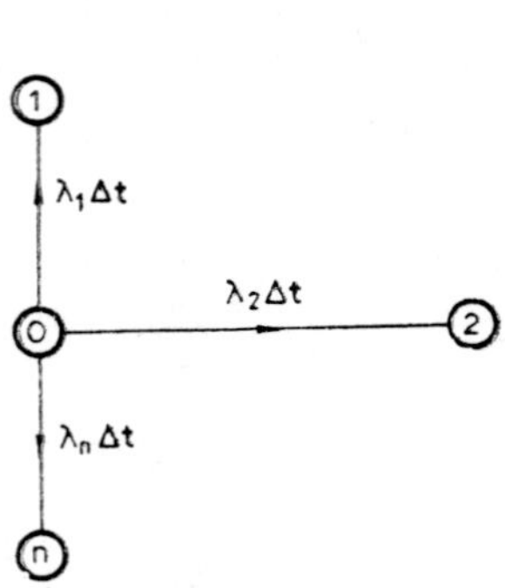

Fig. 6.14. — Markov chain for a non-repairable series system.

The first Markov chain is shown in Fig. 6.14. State "0" is the up state and states "1" to "n" are the failed states corresponding to the component failures. There are no transitions from the failed states, which are absorbing states. The equation of P_0 at $t + \Delta t$ is

$$P_0(t + \Delta t) = P_0(t)(1 - \lambda_s \Delta t) \tag{6.97}$$

where $\lambda_s = \sum_{i=1}^{n} \lambda_i$.

Equation (6.97) is transformed into the differential equation

$$\frac{dP_0(t)}{dt} = - \lambda_s P_0(t). \tag{6.98}$$

The solution of equation (6.97) with initial condition $P_0(0) = 1$ is :

$$P_0(t) = e^{-\lambda_s t}. \tag{6.99}$$

Expression (6.99) is the reliability function of a series system (which has already been obtained by using the logical model).

The second Markov chain model takes into account the transitions from the failed states into state "0". The probabilities of the states at $t + \Delta t$ are :

$$P_i(t + \Delta t) = P_0(t)\lambda_i \Delta t + \\ + P_i(t)(1 - \rho_i \Delta t) \tag{6.100}$$

$$P_0(t) + \sum_{i=1}^{n} P_i(t) = 1.$$

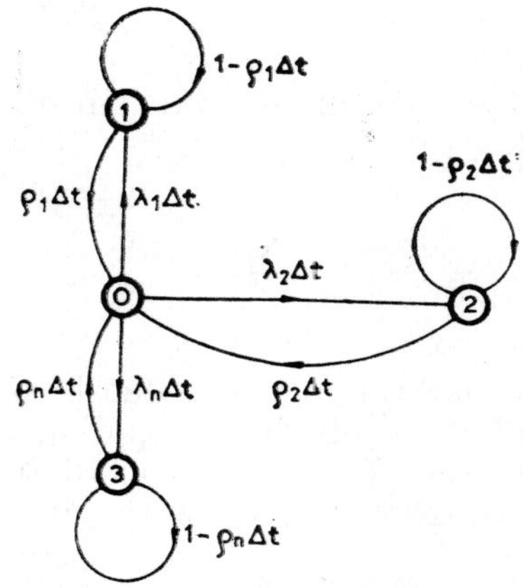

Fig. 6.15. — Markov chain for a repairable series system.

Equations (6.100) are transformed into differential equations

$$\frac{dP_i(t)}{dt} = -\rho_i P_i(t) + \lambda_i P_0(t). \tag{6.101}$$

Taking the Laplace transforms of equations (6.101) with initial conditions $P_i(0) = 0$, we get

$$\sum_{i=1}^{n} P_i^*(s) = \frac{1}{s} - P_0^*(s) = \sum_{i=1}^{n} \frac{\lambda_i}{s + \rho_i} P_0^*(s). \tag{6.102}$$

Finally, the Laplace transform of system availability is

$$A^*(s) = P_0^*(s) = \frac{1}{s\left(1 + \sum\limits_{i=1}^{n} \frac{\lambda_i}{s + \rho_i}\right)}. \tag{6.103}$$

The Laplace transforms of the probabilities associated with the failed states are

$$P_i^*(s) = \frac{\frac{\lambda_i}{s + \rho_i}}{s\left(1 + \sum\limits_{i=1}^{n} \frac{\lambda_i}{s + \rho_i}\right)}. \tag{6.104}$$

The total transition rate of the system into the failed state i is :

$$h_{1i}^*(s) = \lambda_i P_0^*(s) = \frac{\lambda_i}{s\left(1 + \sum\limits_{i=1}^{n} \frac{\lambda_i}{s + \rho_i}\right)}. \tag{6.105}$$

The type 1 renewal density is then :

$$h_1^*(s) = \sum_{i=1}^{n} h_{1i}^*(s) = \frac{\lambda_s}{s\left(1 + \sum_{i=1}^{n} \frac{\lambda_i}{s + \rho_i}\right)}. \tag{6.106}$$

The type 2 renewal density is the total transition rate into state "0" :

$$h_2^*(s) = \sum_{i=1}^{n} \rho_i P_i^*(s) = \frac{\sum_{i=1}^{n} \frac{\lambda_i \rho_i}{s + \rho_i}}{s\left(1 + \sum_{i=1}^{n} \frac{\lambda_i}{s + \rho_i}\right)}. \tag{6.107}$$

From the renewal densities (6.106) and (6.107), the renewal functions may readily be computed, so that predictions of the mean number of system failures are available. Prediction of the next system failure is straightforward, as the interval reliability is :

$$R(t, t + x) = P_0(t)\mathrm{e}^{-\lambda_s x}. \tag{6.108}$$

Consequently, the mean residual life is

$$m(t) = \int_0^{\infty} R(t, t + x)\,\mathrm{d}x = P_0(t) \cdot \frac{1}{\lambda_s} = A(t) \cdot m. \tag{6.109}$$

We shall now derive the asymptotic values of the reliability measures. The availability coefficient is

$$A = \lim_{s \to 0} sA^*(s) = \frac{1}{1 + \sum_{i=1}^{n} \frac{\lambda_i}{\rho_i}} = \frac{\frac{1}{\lambda_s}}{\frac{1}{\lambda_s} + \frac{1}{\lambda_s} \sum_{i=1}^{n} \frac{\lambda_i}{\rho_i}}. \tag{6.110}$$

As it is well known that $A = m/(m + \mu)$, we can derive the system MTTR from relation (6.110) :

$$\mu = \sum_{i=1}^{n} \frac{1}{\rho_i} \frac{\lambda_i}{\lambda_s} = \sum_{i=1}^{n} \mu_i \frac{\lambda_i}{\lambda_s}. \tag{6.111}$$

Relation (6.111) states that the system MTTR is equal to the sum of the MTTRs of its components each multiplied by the probability of the system being in state i, given that it has failed.

The renewal densities are, asymptotically :

$$\lim_{t \to \infty} h_1(t) = \lim_{t \to \infty} h_2(t) = \frac{\lambda_s}{1 + \sum_{i=1}^{n} \frac{\lambda_i}{\rho_i}} = \frac{1}{m + \mu}. \tag{6.112}$$

One is tempted to compute the series system availability by the same reasoning as that employed when the reliability function was computed on the basis of the logical model. As the components are statistically independent, the probability that all components are good is the product of their availabilities, so that in the stationary state one might write:

$$A = \prod_{i=1}^{n} \frac{\rho_i}{\lambda_i + \rho_i} = \frac{1}{\prod_{i=1}^{n}\left(1 + \frac{\lambda_i}{\rho_i}\right)}. \tag{6.113}$$

Comparing relations (6.110) and (6.113), we notice that they give only approximately similar results for $\lambda_i/\rho_i \ll 1$. The non-concordance between relations (6.110) and (6.113) is explained by the fact that unlike the failure processes in $(0, t)$, the state probabilities at t are not independent, being mutually exclusive. Failure of the system in state i prevents the other components following their natural evolution, because the system is not used during its repair. Therefore, it is theoretically wrong to multiply component availabilities in order to compute the availability of a series system. The distinction between independence of the failure processes and the mutual dependence of the resulting state is essential in reliability theory, and is illustrated best in the case of systems with multiple failure modes [38].

We shall now consider the Markov chain associated with a parallel system with two components. As before, the analysis will have two steps. In the first, the system failed state is absorbing, so that the probability of the system being good at time t is its reliability function. In the second step, system recovery is also considered, so the probability of the system being good at t will represent its availability.

If the renewal takes place only upon system failure, the results of the Markov chain analysis will coincide with those obtained on the basis of the logical model. The Markov chain model is particularly useful when preventive renewal is considered, that is the failed element is repaired while the system is still working by virtue of the second element. This is precisely a continuous replacement policy, preventive renewal being carried out when the system reaches the worn state (with one element failed, the other working). Reliability analysis under CRP may be performed only on the basis of the Markov chain model.

We assume that the parallel system has two identical CFR components with failure rates λ and repair rates ρ. The graph of the first Markov chain model is depicted in Fig. 6.16. In state "0" both components are working. By the failure of any component, the system reaches the worn state "1", where one component is still working while the other undergoes repair. If the working component fails before the repair of the other component has been completed, the system reaches the failed state "2". State "2" is absorbing in the first model as the reliability function should be derived at this step.

The equations of the state probabilities are :

$$\begin{aligned} P_0(t+\Delta t) &= P_0(t)(1-2\lambda\Delta t)+P_1(t)\rho\Delta t \\ P_0(t+\Delta t) &= P_0(t)\,2\lambda\Delta t+P_1(t)[1-(\lambda+\rho)]\Delta t. \end{aligned} \tag{6.114}$$

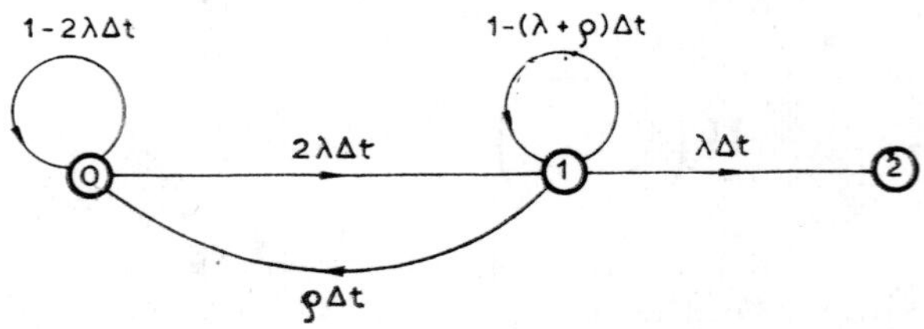

Fig. 6.16.—Markov chain model for reliability analysis of a parallel system.

A transformation of equations (6.114) into differential form yields :

$$\begin{aligned} \frac{\mathrm{d}P_0(t)}{\mathrm{d}t} &= -\,2\lambda P_0(t)+\rho P_1(t) \\ \frac{\mathrm{d}P_1(t)}{\mathrm{d}t} &= 2\lambda P_0(t)-(\lambda+\rho)P_1(t). \end{aligned} \tag{6.115}$$

We shall apply the Laplace transform to equations (6.115) considering the system to be initially in state "0", that is $P_0(0)=1$ and $P_1(0)=0$. We denote the state probabilities by P_{00} and P_{10} when the initial state is "0".

$$\begin{aligned} (s+2\lambda)P^*_{00}(s)-\rho P^*_{10}(s) &= 1 \\ -\,2\lambda P^*_{00}(s)+(s+\lambda+\rho)P^*_{10} &= 0. \end{aligned} \tag{6.116}$$

Solving the algebraic equations (6.116), gives the Laplace transform of the system reliability function, which is equal to the sum of the probabilities of the "good" states.

$$R^*_{0}(s)=P^*_{00}(s)+P^*_{10}(s)=\frac{s+3\lambda+\rho}{s^2+(3\lambda+\rho)s+2\lambda^2}. \tag{6.117}$$

From expression (6.117), the system MTTF may be readily derived :

$$m_0=\int_0^\infty R_0(t)\,\mathrm{d}t=R^*_{0}(s)|_{s=0}=\frac{3\lambda+\rho}{2\lambda^2}. \tag{6.118}$$

The original of the reliability function is easy to find, as its Laplace transform (6.117) is a rational function.

We shall now take the Laplace transform of equations (6.115) in the case when the system is initially in state "1", that is $P_0(0)=0$ and

$P_{01}(0) = 1$. We denote the state probabilities by P_{01} and P_{11} under this initial condition.

$$\begin{aligned} (s + 2\lambda)P_{01}^*(s) - \rho P_{11}^*(s) &= 0 \\ - 2\lambda P_{01}^*(s) + (s + \lambda + \rho)P_{11}^*(s) &= 1. \end{aligned} \tag{6.119}$$

Solving equations (6.119), we get the reliability function when the initial system state is "1"

$$R_1^*(s) = P_{01}^*(s) + P_{11}^*(s) = \frac{s + 2\lambda + \rho}{s^2 + (3\lambda + \rho)s + 2\lambda^2}. \tag{6.120}$$

The system MTTF in this case is :

$$m_1 = \int_0^\infty R_1(t)\, dt = R_1^*(s)\,|_{s=0} = \frac{2\lambda + \rho}{2\lambda^2}. \tag{6.121}$$

We next proceed to the second step of the analysis, when system renewal is also considered. After failure, the system may recover either to state "0" or to state "1". If one waits until the repair of both components is completed, the system recovers to state "0". Conversely, if operation of the system starts as soon as the repair of one component is completed, the system recovers to state "1". In both situations, one must distinguish between the two server case (when repair of the two components may be performed simultaneously) and the one server case (when the second component to fail must wait for the repair of the other component to be completed).

Consequently, there are four cases, each with a different distribution law of the time for system repair. We denote by $g_{ij}(t)$ the pdf of the time-to-repair of the system and by μ_{ij} the system MTTR when it recovers to state i with j servers. It is obvious that

$$\begin{aligned} g_{11}(t) &= \rho e^{-\rho t}; \qquad \mu_{11} = \frac{1}{\rho} \\ g_{12}(t) &= 2\rho e^{-2\rho t}; \qquad \mu_{12} = \frac{1}{2\rho}. \end{aligned} \tag{6.122}$$

When the system recovers to state "0" with one server, the system time-to-repair is the sum of the times-to-repair of the two components. Consequently, the pdf of the system time-to-repair is the convolution of two exponentials

$$g_{01}(t) = \rho e^{-\rho t} \otimes \rho e^{-\rho t} = \rho^2 t e^{-\rho t}; \quad \mu_{01} = \frac{2}{\rho}. \tag{6.123}$$

When the system recovers to state "0" with two servers, the cdf of the time-to-repair is $G_{02}(t) = (1 - e^{-\rho t})^2$, so that

$$g_{02}(t) = 2\rho e^{-\rho t}(1 - e^{-\rho t}); \quad \mu_{02} = \frac{2}{\rho} - \frac{1}{2\rho} = \frac{3}{2\rho}. \tag{6.124}$$

The four cases examined above correspond to four alternating renewal processes (Figure 6.17). The alternating processes corresponding to the cases when the system recovers to state "0" are simple, as all times-to-failure are distributed according to $f_0(t) = -R_0'(t)$, with means equal to m_0, and all times-to-repair are distributed according to $g_{01}(t)$ or $g_{02}(t)$ with means equal to μ_{01} or μ_{02} for one or two servers respectively.

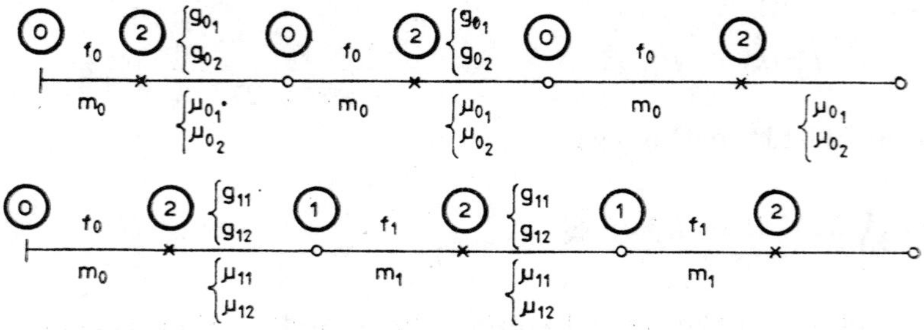

Fig. 6.17. — Alternating renewal processes for a parallel system.

The alternating processes corresponding to the cases when the system recovers to state "1" are general, as the first time-to-failure is distributed according to $f_0(t)$ with mean m_0, while the other times-to-failure are distributed according to $f_1(t) = -R_1'(t)$ with means equal to m_1. Thus, m_0 is the MTTFF and m_1 is the MTBF of the system. The times-to-repair are distributed according to $g_{11}(t)$ or $g_{12}(t)$ with means equal to μ_{11} or μ_{12}, for one or two servers respectively. The theory of alternating renewal processes, developed in Chapter 5, may be used to derive reliability measures for the parallel repairable system.

We shall consider only the case when the system recovers to state "1" with two servers, and derive the reliability measures for the stationary state of the process.

The asymptotical renewal densities are :

$$\lim_{t\to\infty} h_1(t) = \lim_{t\to\infty} h_2(t) = \frac{1}{m_1 + \mu_{12}} = \frac{1}{\dfrac{2\lambda + \rho}{2\lambda^2} + \dfrac{1}{2\rho}} =$$

$$= \frac{2\lambda^2\rho}{(\lambda + \rho)^2}. \tag{6.125}$$

The mean number of failures in an interval $(t, t + x)$ when $t \to \infty$ is then :

$$\lim_{t\to\infty} H_1(t, t + x) = \frac{2\lambda^2\rho x}{(\lambda + \rho)^2} \tag{6.126}$$

and is equal to the mean number of repairs. The interval reliability function (see Chapter 5) is :

$$\lim_{t\to\infty} R(t, t + x) = \frac{1}{m_1 + \mu_{12}} \int_x^\infty R_1(t)\, dt. \tag{6.127}$$

It is more convenient to consider the Laplace transform of expression (6.127):

$$\mathscr{L} \lim_{t\to\infty} R(t, t+x) = \frac{m_1}{s(m_1+\mu_{12})} - \frac{R_1^*(s)}{s(m_1+\mu_{12})} =$$

$$= \frac{A}{s}\left(1 - \frac{R_1^*(s)}{m_1}\right) = \frac{(\rho^2+2\lambda\rho)s+(4\lambda^2+5\lambda\rho+\rho^2)\rho}{(\lambda+\rho)^2[s^2+(3\lambda+\rho)s+2\lambda^2]}. \quad (6.128)$$

The mean residual life is obtained directly from expression (6.128):

$$\lim_{t\to\infty} m(t) = \mathscr{L} \lim_{t\to\infty} R(t, t+x)\big|_{s=0} = \frac{(4\lambda^2+5\lambda\rho+\rho^2)\rho}{2\lambda^2(\lambda+\rho)^2}. \quad (6.129)$$

Finally, the availability coefficient is

$$A = \frac{m_1}{m_1+\mu_{12}} = \frac{(2\lambda+\rho)\rho}{(\lambda+\rho)^2}. \quad (6.130)$$

Derivation of the above reliability measures in the transitory state of the alternative process is more cumbersome, but may in principle be performed according to equations:

$$h_1^*(s) = \frac{f_0^*(s)}{1 - f_1^*(s)g_{12}^*(s)};$$

$$h_2^*(s) = \frac{f_0^*(s)\, g_{12}^*(s)}{1 - f_1^*(s)g_{12}^*(s)} \quad (6.131)$$

$$H_1^*(s) = \frac{h_1^*(s)}{s}; \quad H_2^*(s) = \frac{h_2^*(s)}{s};$$

$$A^*(s) = R_0^*(s) + R_1^*(s) \cdot h_2^*(s).$$

The other three cases may be treated similarly, following the theory of alternating renewal processes. It should be noted that this method quite easily yields the reliability measures in the stationary state of the renewal process, but is difficult to apply in its transitory state.

We shall now derive the same reliability measures for the repairable parallel system, using the Markov chain model. Figure 6.18 depicts this model for the case when the system recovers to state "1". The transition rate from state "2" to state "1" is ρ or 2ρ, for one or two servers respectively. For the two-server case, the state probabilities are:

$$P_0(t+\Delta t) = P_0(t)(1 - 2\lambda\Delta t) + P_1(t)\rho\Delta t$$
$$P_1(t+\Delta t) = P_0(t)\, 2\lambda\Delta t + P_1(t)[1-(\lambda+\rho)\Delta t] + P_2(t)\, 2\rho\Delta t. \quad (6.132)$$

Transforming equations (6.132) into differential equations, and noting that $P_0(t) + P_1(t) + P_2(t) = 1$, we get :

$$\frac{dP_0(t)}{dt} = -2\lambda P_0(t) + \rho P_1(t)$$
$$\frac{dP_1(t)}{dt} = (2\lambda - 2\rho)P_0(t) - (\lambda + 3\rho)P_1(t) + 2\rho. \tag{6.133}$$

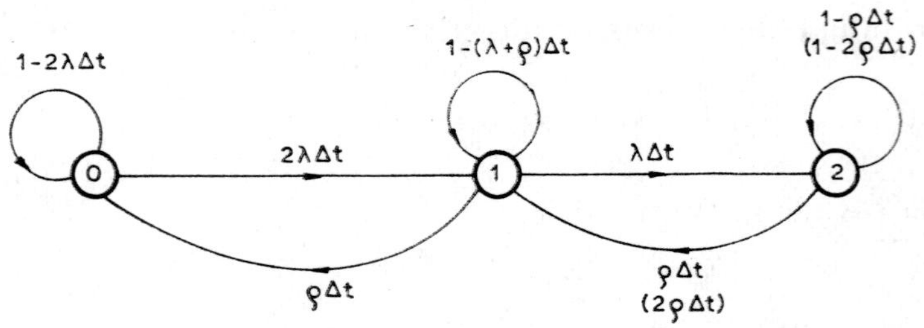

Fig. 6.18. — Markov chain model for availability analysis of a parallel system which recovers to the worn state.

The Laplace transform of equations (6.133), with initial conditions $P_0(0) = 1$ and $P_1(0) = 0$, are

$$(s + 2\lambda)P_0^*(s) - P_1^*(s) = 1$$
$$(2\rho - 2\lambda)P_0^*(s) + (s + \lambda + 3\rho)P_1^*(s) = \frac{2\rho}{s}. \tag{6.134}$$

The solutions of equations (6.134) are :

$$P_0^*(s) = \frac{s^2 + (\lambda + 3\rho)s + 2\rho^2}{s[s^2 + 3(\lambda + \rho)s + 2(\lambda + \rho)^2]}$$
$$P_1^*(s) = \frac{2\lambda(s + 2\rho)}{s[s^2 + 3(\lambda + \rho)s + 2(\lambda + \rho)^2]}. \tag{6.135}$$

The availability is obtained by summing the probabilities (6.135) :

$$A^*(s) = \frac{s^2 + 3(\lambda + \rho)s + 2\rho(\rho + 2\lambda)}{s(s + \lambda + \rho)(s + 2\lambda + 2\rho)}. \tag{6.136}$$

The availability coefficient is :

$$\lim_{t\to\infty} A(t) = \lim_{s\to 0} sA^*(s) = \frac{\rho(2\lambda + \rho)}{(\lambda + \rho)^2} \tag{6.137}$$

which coincides with expression (6.129), obtained by the alternating process method.

The original of the avilability expression (6.136) is :

$$A(t) = \frac{\rho(2\lambda + \rho)}{(\lambda + \rho)^2} + \frac{2\lambda^2}{(\lambda + \rho)^2} e^{-(\lambda+\rho)t} - \frac{\lambda^2}{(\lambda + \rho)^2} e^{-2(\lambda+\rho)t}. \quad (6.138)$$

The renewal densities may also be derived by the Markov chain method. The type I renewal density is the total probability that the system enters failed state "2", and the type II renewal density is the total probability that the system recovers to state "1". We have

$$h_1^*(s) = P_1^*(s) \cdot \lambda = \frac{2\lambda^2(s + 2\rho)}{s(s + \lambda + \rho)(s + 2\lambda + 2\rho)};$$

$$h_2(s) = P_2^*(s) \cdot 2\rho = \left[\frac{1}{s} - P_0^*(s) - P_1^*(s)\right] 2\rho = \quad (6.139)$$

$$= \frac{4\lambda^2\rho}{s(s + \lambda + \rho)(s + 2\lambda + 2\rho)}.$$

The renewal functions may be readily derived from expressions (6.139), so the mean number of failure may be computed both in the transitory and the stationary states of the renewal process.

Once the type II renewal density is known, the interval reliability and the MRL may be computed from the relations derived on the basis of the alternating process model. The Markov chain analysis has the advantage of providing more easily closed forms for the reliability measures of a repairable parallel system in the transitory state of the renewal process. However, the model is restricted to systems with components which have their times-to-failure and times-to-repair exponentially distributed.

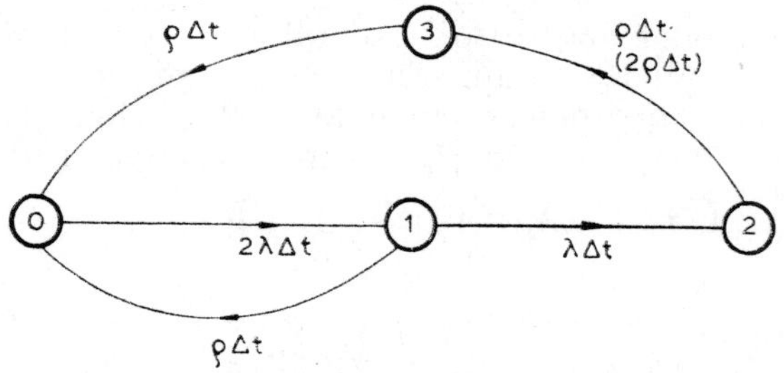

Fig. 6.19. — Markov chain model for availability analysis of a parallel system which recovers to its initial state.

The Markov chain may also be used in analysing the cases when the system recovers to state "0" with one or two servers. A difficulty arises here, since the transition rate between state "2" and state "0" is by no means constant. To circumvent this difficulty, system state "3" is considered, in which the repair of only one component has been completed. The graph is presented in Fig. 6.19. The transition rate from state "2" to state "3" is ρ in the one-server case and 2ρ in the two-server case. A quantitative analysis may be made in a way similar to the case already described in detail.

In the case of a parallel system with two servers, a simple relation between the system availability and the component availabilities may be derived. The component availabilities are given by relation (6.95), that is:

$$A_1(t) = A_2(t) = \frac{\rho}{\lambda + \rho} + \frac{\lambda}{\lambda + \rho} e^{-(\lambda+\rho)t}. \tag{6.140}$$

Expressions (6.138) and (6.140) satisfy the relation

$$A(t) = A_1(t) + A_2(t) - A_1(t) \cdot A_2(t). \tag{6.141}$$

Relation (6.141), which is similar to the reliability expression for the parallel system, could have been derived on the basis of the logical model by the following reasoning. The parallel system is in a good state when either of the two components is good. If the component states are statistically independent, relation (6.141) is directly derived from the formula for computing the probability of a reunion of events.

The component states are independent only because two servers were considered. Thus, one component does not have to wait for the repair of the other to be completed, so the component renewal processes are mutually independent, just like their failure processes.

Consequently, relation (6.141) is true for parallel systems with enough repair resources to assure the independence of the component renewal processes. However, it may be used even in cases when the number of servers is less than the number of components, since it yields approximate results with a fair accuracy.

The logical model may also be employed to compute the renewal densities of the parallel system. The system enters state "2" around instant t when one component is good at t, the other has already failed and the good component fails in $(t, t + \Delta t)$. We may then write

$$h_1(t) = \lambda A_1(t)\bar{A}_2(t) + \lambda \bar{A}_1(t) A_2(t) = 2\lambda A\bar{A} =$$

$$= 2\lambda \left(\frac{\rho}{\lambda + \rho} + \frac{\lambda}{\lambda + \rho} e^{-(\lambda+\rho)t} \right) \cdot \left(\frac{\lambda}{\lambda + \rho} - \frac{\lambda}{\lambda + \rho} e^{-(\lambda+\rho)t} \right) =$$

$$= \frac{2\lambda^2}{(\lambda + \rho)^2} [\rho - (\rho - \lambda) e^{-(\lambda+\rho)t} - \lambda e^{-2(\lambda+\rho)t}]. \tag{6.142}$$

Similarly, the system enters state "1" around t when both components have already failed at t and repair of one is completed in $(t, t + \Delta t)$. The type II renewal density is then

$$h_2(t) = 2\rho \bar{A}_1 \bar{A}_2 = 2\rho \bar{A}^2 = \frac{2\lambda^2 \rho}{(\lambda + \rho)^2} [1 - e^{-(\lambda+\rho)t}]^2. \tag{6.143}$$

Expressions (6.142) and (6.143) are precisely the originals of expressions (6.139) obtained by the Markov chain model.

The possibility of performing an availability analysis of the parallel system directly on the basis of the logical model shows that the Markov chain model is essential only for deriving the reliability function under the continuous replacement policy. Even this step of the reliability analysis may be circumvented if the reliability function is approximated according to the relation (see Section 5.3):

$$R(t) \simeq \exp\left[-\int_0^t \Lambda(u)\,\mathrm{d}u\right] \tag{6.144}$$

where $\Lambda(u) = \dfrac{h_1(t)}{A(t)}$.

The system availability may be computed according to relation (6.141), yielding expression (6.138) and the type I renewal density may be computed from relation (6.142). Consequently, the reliability function may be approximated by relation (6.144) where:

$$\Lambda(t) = \frac{2\lambda^2[\rho - (\rho - \lambda)\mathrm{e}^{-(\lambda+\rho)t} - \lambda\mathrm{e}^{-2(\lambda+\rho)t}]}{\rho(2\lambda + \rho) + 2\lambda^2\mathrm{e}^{-(\lambda+\rho)t} - \lambda^2\mathrm{e}^{-2(\lambda+\rho)t}}\,. \tag{6.145}$$

As expression (6.145) seems rather complicated, we give a numerical example [39]. Let $\lambda = 2$ and $\rho = 3$. From relation (6.117) we compute the true reliability function, given that the system is initially in state "0":

$$R_0^*(s) = \frac{s+9}{s^2+9s+8} = \frac{s+9}{(s+8)(s+1)}\,;$$
$$R_0(t) = \frac{8}{7}\,\mathrm{e}^{-t} - \frac{1}{7}\,\mathrm{e}^{-8t}. \tag{6.146}$$

The specific value of $\Lambda(t)$ is:

$$\Lambda(t) = \frac{8(3 - \mathrm{e}^{-5t} - 2\mathrm{e}^{-10t})}{21 + 8\mathrm{e}^{-5t} - 4\mathrm{e}^{-10t}} = \frac{8(1 - \mathrm{e}^{-5t})}{7 - 2\mathrm{e}^{-5t}}\,. \tag{6.147}$$

The approximate value of the reliability function is, according to relation (6.144):

$$R_0(t) \simeq \exp\left[-\int_0^t \frac{8(1 - \mathrm{e}^{-5u})}{7 - 2\mathrm{e}^{-5u}}\,\mathrm{d}u\right] = \left(\frac{7}{5}\,\mathrm{e}^{-2t} - \frac{2}{5}\,\mathrm{e}^{-7t}\right)^{\frac{4}{7}}. \tag{6.148}$$

A graphical representation of expressions (6.146) and (6.148) indicates the accuracy of the approximation.

We shall now examine the application of the Markov chain model to a general system with structure function $\varphi(x_1, x_2, \ldots, x_n)$.

In the first step, the Markov chain model is derived by state enumeration and by specifying the transition rates, without taking into account the transitions backwards from the failed states. The differential equations of the state probabilities are then solved, yielding the system reliability function.

In the second step, alternating renewal process theory is applied to compute the availability and the renewal functions in the stationary state of the system renewal process. As the alternating renewal process theory is difficult to apply in the transitory state, the Markov chain model should be used in this case. The transitions from the failed states are incorporated in the model and the new differential equations of the state probabilities solved, yielding the availability and the renewal densities.

Solving the state differential equations is an arduous task even for moderately complex systems, so an approximate method was developed to derive system reliability by the Markov chain model [40]. Let w_k be a trajectory, that is, a succession of system states $(s_{k_0}, s_{k_1}, \ldots, s_{kn_k})$, through which the system passes from the initial state s_{k_0} to a failed state s_{kn_k}. The duration of a trajectory is a random variable with cdf $F(t/w_k)$. The probability of system failure in $(0, t)$ is, by the total probability formula,

$$F(t) = \sum_k F(t/w_k)P(w_k). \tag{6.149}$$

When a continuous replacement policy is in use, the number of trajectories is infinite, since the system may repeatedly revisit the same state by component repair. Consequently, expression (6.149) may be computed by simulation. N trajectories $\{w_k;\ k = 1, 2, \ldots, N\}$ are generated by Monte Carlo methods, according to their probabilities $\{P(w_k);\ k = 1, 2, \ldots, N\}$. The system failure probability is estimated by the arithmetic mean of the cdfs associated with the durations of the simulated trajectories

$$\hat{F}(t) = \frac{\sum_{k=1}^{N} F(t/w_k)}{N}. \tag{6.150}$$

To show how a trajectory w_k is simulated, we consider the state $S_{k,i-1} \in w_k$ and the transition rates $\gamma_{i_1}, \gamma_{i_2}, \ldots, \gamma_{i_{m_i}}$ from state $S_{k,i-1}$ to the states $A_1, A_2, \ldots, A_{m_i}$ which are accessible from state $S_{k,i-1}$. Any of the states $A_1, A_2, \ldots, A_{m_i}$ may become state S_{ki} of the sequence w_k. The probability that the system passes into state A_j, given that it has left state $S_{k,i-1}$, is γ_{ij}/Γ_i, where $\Gamma_i = \sum_{j=1}^{m_i} \gamma_{ij}$. To build up a certain sequence of states, interval $(0, 1)$ is divided into m_i subintervals of lengths $\{\gamma_{ij}/\Gamma_i;\ j=1, 2, \ldots, m_i\}$ respectively, as in Fig. 6.20.

A random number is drawn from a Monte Carlo table, which contains specific values of a random variable uniformly distributed in (0, 1). If the random number belongs to subinterval γ_{ij}/Γ_i, state A_j becomes state S_{k_i} in the sequence w_k.

Figure 6.20 shows that the random number belongs to subinterval γ_{i_2}/Γ_i, so that state A_2 is state S_{k_i} of the sequence w_k. To complete the sequence, the procedure is repeated until a failed state S_{kn_k} is reached.

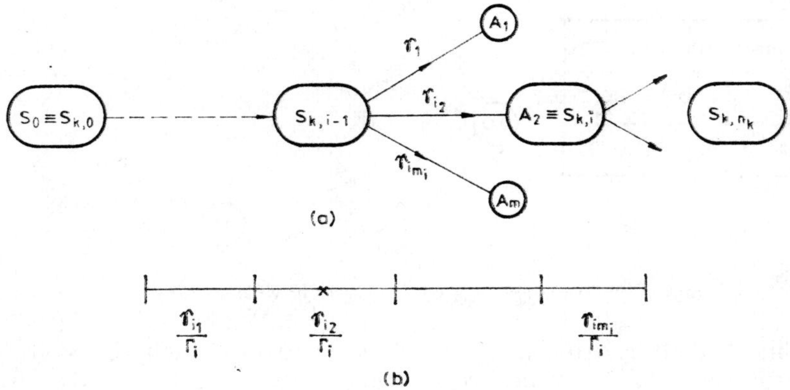

Fig. 6.20. — Trajectory simulation.

It is obvious that the sequences are not mutually disjoint, as they have in common at least the initial state S_0.

The time for which the system remains in state $S_{k,i-1}$ is a random variable exponentially distributed with parameter Γ_i. Consequently, the duration of a sequence w_k is distributed according to a convolution of n_k exponential distributions with parameters $\{\Gamma_i;\ i = 1, 2, \ldots, n_k\}$. Since the Laplace transform of an exponential pdf is $\Gamma_i/(s + \Gamma_i)$, the Laplace transform of the cdf associated with the duration of a sequence is

$$F^*(s/w_k) = \frac{1}{s} \prod_{i=1}^{n_k} \frac{\Gamma_i}{s + \Gamma_i} \,. \tag{6.151}$$

Taking the arithmetic mean of the cdfs (6.151) according to relation (6.150), we get an estimate of the system failure probability in $(0, t)$, which completes the analysis.

We shall consider, as an example, a parallel system with two dissimilar components which have constant failure and repair rates. The Markov chain model of the system is given in Fig. 6.21, together with the specific values of the failure and repair rates.

The system is initially in state (0, 0) when both components are good. If component 1 fails, the system passes into state (1, 0) with transition rate λ_1, while if component 2 fails, the system passes into state (0, 1) with transition rate λ_2. The probabilities that the system passes into state (1, 0) or into state (0, 1), given that it has left state (0, 0) are respectively $\lambda_1(\lambda_1 + \lambda_2) = 0.4$ and $\lambda_2/(\lambda_1 + \lambda_2) = 0.6$. From state (1, 0), the system may pass into the failed state (1, 1) by failure of component 2 or into the initial state (0,0) by repair of component 1. The probabilities associated with these transitions, conditioned by the system leaving state (1, 0) are $\lambda_2(\lambda_2 + \rho_1) = 0.23$ and $\rho_1(\lambda_2 + \rho_1) = 0.77$, respectively.

Component	$\lambda(h^{-1})$	$\rho(h^{-1})$
1	$2 \cdot 10^{-5}$	10^{-4}
2	$3 \cdot 10^{-5}$	$2 \cdot 10^{-4}$

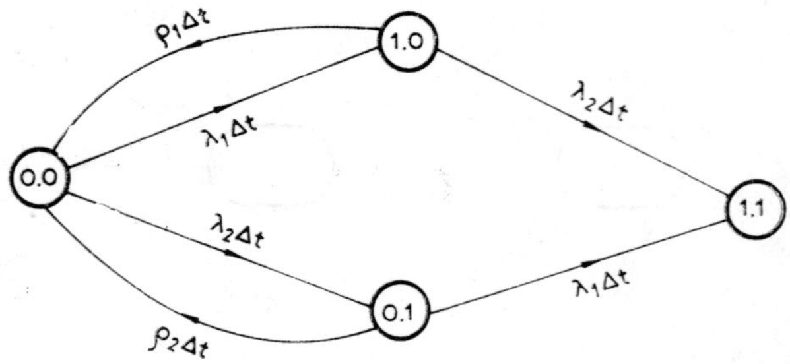

Fig. 6.21. — Markov chain model for a parallel system with two dissimilar components.

The last alternative is in state (0, 1), from which the system may pass into state (1, 1) by failure of component 1 or into state (0, 0) by repair of component 2, whichever occurs first. The conditional probabilities of these events are $\lambda_1(\lambda_1 + \rho_2) = 0.09$ and $\rho_2(\lambda_1 + \rho_2) = 0.91$, respectively.

Possible trajectories from state (0, 0) to state (1, 1) are now generated using a table of random numbers and Fig. 6.22.

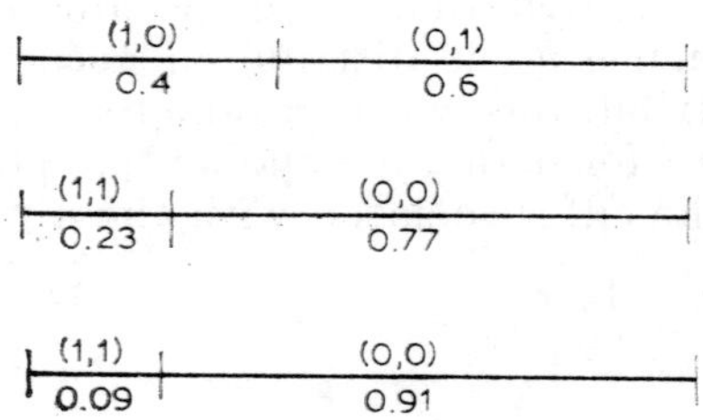

Fig. 6.22. — Generation of the next state in a sequence:
(a) from state (0,0); (b) from state (1,0); (c) from state (0,1).

Five generated trajectories are given in Table 6.3. Let us consider the second trajectory. It begins in state $S_{20} = (0, 0)$. The random number 0.48 being greater than 0.4, the next state of the sequence is $S_{21} = (0, 1)$. The next random number drawn in the Monte Carlo table is 0.05, which is less than 0.09, so the next system state is $S_{22} = (1, 1)$, which is the

Table 6.3

Generating trajectories for a parallel system

Random numbers	0.72	0.58	0.79	0.30	0.00	0.89	0.68	0.87	0.84	0.16	0.27	0.58
States	(0,1)	(0,0)	(0,1)	(0,0)	(1,0)	(0,0)	(0,1)	(0,0)	(0,1)	(0,0)	(1,0)	(0,0)
Sequences				I (First trajectory)					→			
Random numbers	0.99	0.95	0.28	0.14	0.48	0.05	0.09	0.61	0.52	0.45	0.24	0.24
States	(0,1)	(0,0)	(1,0)	(1,1)	(0,1)	(1,1)	(1,0)	(0,0)	(0,1)	(0,0)	(1,0)	(0,0)
Sequences						II →			III		→	
Random numbers	0.99	0.33	0.34	0.68	0.39	0.35	0.79	0.13	0.09	0.04	0.10	0.45
States	(0,1)	(0,0)	(1,0)	(0,0)	(1,0)	(0,0)	(0,1)	(0,0)	(1,0)	(1,1)	(1,0)	(0,0)
Sequences										→	IV	→
Random numbers	0.42	0.07	0.77	0.57	0.76	0.77	0.39	0.75	0.26	0.27	0.25	0.66
States	(0,1)	(1,1)	(0,1)	(0,0)	(0,1)	(0,0)	(1,0)	(0,0)	(1,0)	(0,0)	(1,0)	(0,0)
Sequences	→				V			→				
Random numbers	0.64	0.47	0.25	0.73	0.13	0.75	0.25	0.16	0.28	0.76	0.61	0.81
States	(0,1)	(0,0)	(1,0)	(0,0)	(1,0)	(0,0)	(1,0)	(1,1)				
Sequences								→				

failed state, so the sequence is completed. The other trajectories are similarly generated. Their great lengths are explained by the existence of cycles.

We shall now derive the cdf of the duration of the second trajectory. The duration is the sum of the times the system spends in states (0, 0) and (0, 1). Both times are exponentially distributed, with parameters $\lambda_1 + \lambda_2$ and $\lambda_1 + \rho_2$. The Laplace transform of the cdf is, by relation (6.151) :

$$F^*(s/w_2) = \frac{1}{s} \frac{(\lambda_1 + \lambda_2)(\lambda_1 + \rho_2)}{(s + \lambda_1 + \lambda_2)(s + \lambda_1 + \rho_2)}. \qquad (6.152)$$

The other cdfs are similarly computed. For the fourth trajectory, we have :

$$F^*(s/w_4) = \frac{1}{s} \frac{(\lambda_1 + \lambda_2)^2(\lambda_1 + \rho_2)(\lambda_2 + \rho_1)}{(s + \lambda_1 + \lambda_2)^2(s + \lambda_1 + \rho_2)(s + \lambda_2 + \rho_1)}. \qquad (6.153)$$

The long trajectories (more than 10 states in a sequence) have small failure probabilities since it is highly unlikely that any of them will be completed during the system operation period. Therefore their contributions may be neglected when the system failure probability is computed. We shall thus neglect in our example the cdfs associated with trajectories I, III and V. The Laplace transform of the system failure probability then becomes

$$\hat{F}^*(s) = \frac{1}{5} [F^*(s/w_2) + F^*(s/w_4)] =$$

$$= \frac{1}{5s} \frac{(\lambda_1 + \lambda_2)(\lambda_1 + \rho_2)}{(s+\lambda_1+\lambda_2)(s+\lambda_1+\rho_2)} \left[1 + \frac{(\lambda_1 + \lambda_2)(\lambda_2 + \rho_1)}{(s+\lambda_1+\lambda_2)(s+\lambda_2+\rho_1)}\right]. \qquad (6.154)$$

The estimate (6.154) is not very accurate, since very few trajectories have been simulated. In a computer-aided analysis, one would simulate at least 100 trajectories, but the algorithm remains the same.

The Markov chain model is useful for an exact analysis of simple systems or an approximate analysis of complex systems. However, its application is restricted to system with CFR components and exponentially distributed component repair times. More general Markov chain models, able to deal with time-varying transition rates,have been developed for certain specific systems [41, 42, 43].

6.4 The fault tree model

The fault tree is a structural model which shows the unfavourable situation arising when the system performances are beyond the specified tolerance intervals. As a matter of fact, any reliability analysis should account for the circumstances of system failure. The logical model is oriented towards the assurance of satisfactory system performance, and shows how the system components contribute to it. Such a model is useful when designs for system structure are developed, whereas the fault tree model is more appropriate to the critical point of view of the reliability engineer drawing attention to the weak links of a proposed structure.

To define a fault tree, the equations of the functional model are considered :

$$y_i = f_i(\gamma_1^{(1)}, \gamma_1^{(2)}, \ldots, \gamma_1^{(l_1)}, \gamma_2^{(1)}, \gamma_2^{(2)}, \ldots, \gamma_2^{(l_2)}, \ldots, \gamma_n^{(1)}, \gamma_n^{(2)}, \ldots,$$
$$\ldots, \gamma_n^{(l_n)}); \; i = 1, 2, \ldots, p. \tag{6.155}$$

The fault tree model, just like the logical model, is obtained from the functional equations by reducing the dimensions of the output and state vectors. Thus instead of the p output variables, a binary variable $S(t)$ is defined so that

$$\begin{aligned} S(t) &= 0 \quad \text{if } \forall\, i = 1, 2, \ldots, p : y_{i_L} \leqslant y_i \leqslant y_{i_U} \\ S(t) &= 1 \quad \text{if } \exists\, i : y_i < y_{i_L} \text{ or } y_i > y_{i_U}. \end{aligned} \tag{6.156}$$

Similarly a binary variable x_j is defined for each component so that

$$\begin{aligned} x_j(t) &= 0 \text{ if } \forall\, k = 1, 2, \ldots, l_i \;\; \gamma_{j_L}^{(k)} \leqslant \gamma_j^{(k)} \geqslant \gamma_{j_U}^{(k)} \\ x_j(t) &= 1 \text{ if } \exists\, k : \gamma_j^{(k)} < \gamma_{j_L}^{(k)} \text{ or } \gamma_j^{(k)} > \gamma_{j_U}^{(k)}. \end{aligned} \tag{6.157}$$

The state-output equation may then be expressed by the boolean structure function

$$S = \varphi(x_1, x_2, \ldots, x_n). \tag{6.158}$$

The structure function (6.158) defines the fault tree as a dual of the logical model.

A useful representation of the fault tree is a combinational circuit with inputs $\{x_j ; j = 1, 2, \ldots, n\}$ and output S. When an input becomes 1, a primary event occurs. When the output value is 1 the TOP event occurs (system failure). A fault tree with only AND and OR gates describes a coherent system. The presence of NOT or EXCLUSIVE OR gates implies system non-coherence. The symbolism used in the fault tree is represented in Fig. 6.23. The event representations are as follows [44].

(a) The rectangle identifies an event that results from the combinations of fault events through the input logic gate.

(b) The circle describes a basic fault event that needs no further development (a primary event). The frequency and failure mode of items thus identified are derived from empirical data.

(c) The triangles are used as transfer symbols. A line from the apex of the triangle indicates a transfer in, and a line from the side or bottom indicates a transfer out.
(d) The diamond describes a fault event that is considered basic in a given fault tree. The possible causes of the event are not developed further, because the event is not of sufficient importance or the necessary information is unavailable.

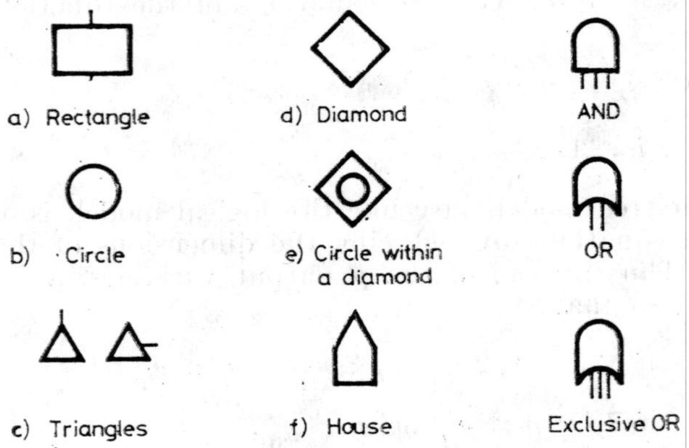

Fig. 6.23. — Fault tree symbolism.

(e) The circle within a diamond indicates a subtree exists, but that subtree was evaluated separately and its quantitative results inserted as though it were a component.
f) The house is used as a switch to include or eliminate parts of the fault tree, as those parts may or may not apply to certain situations.

The fault tree model is adequate for an exact analysis of systems without renewal and an approximate analysis of renewal systems.
When there are no renewals, the goal of the analysis is to obtain a relationship between the system failure probability F_S and the component failure probabilities $F_1, F_2, \ldots, F_n$. Such a relationship may be derived from the structure function (6.158) by any of the methods described in Section 6.2. To this end, the structure function is written in the form

$$S = \bigcup_{i=1}^{m} x_{i_1} x_{i_2} \ldots x_{i_{n_i}} = \bigcup_{i=1}^{m} S_i. \tag{6.159}$$

Each term of expression (6.159) is a minimal cut set, that is, a minimal set of components which, by their failure, cause the system to fail. Once the minimal cut sets are specified, one may apply the methods of total probability inclusion exclusion, sum of disjoint products, the algebraic method and so on, with the only difference that the probability of a boolean variable being 1 signifies here a probability of failure, not a reliability function.

We shall consider as an example the communication system described in Section 6.2. The cut sets may be derived by inspection, so that the structure function of the fault tree is

$$S = (x_1 \cap x_2 \cap x_5) \cup (x_1 \cap x_4 \cap x_6) \cup (x_1 \cap x_2 \cap x_3 \cap x_4) \cup$$
$$\cup (x_1 \cap x_3 \cap x_5 \cap x_6) \tag{6.160}$$

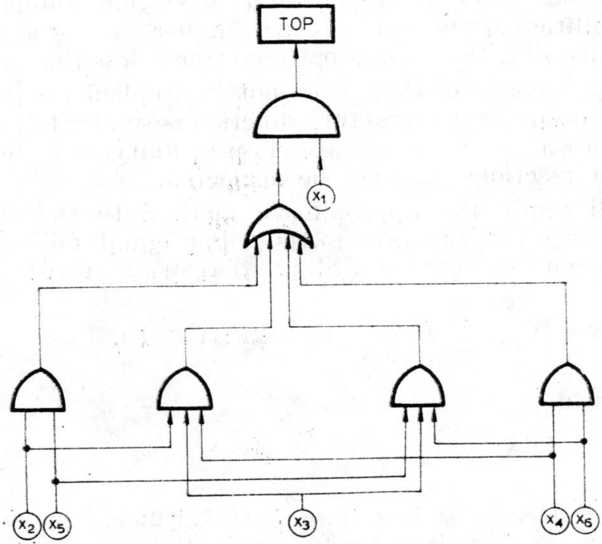

Fig. 6.24. — Fault tree of a communication system.

The combinational circuit associated with the structure function (6.160) is depicted in Fig. 6.24.

We shall apply the algebraic method to derive the system failure probability. Using the logical-algebraic equivalences, the logical function (6.160) is transformed into the algebraic function

$$S = x_1x_2x_5 + x_1x_4x_6 + x_1x_2x_3x_4 + x_1x_3x_5x_6 - x_1x_2x_3x_4x_5 -$$
$$- x_1x_2x_3x_5x_6 - x_1x_2x_4x_5x_6 - x_1x_2x_3x_4x_6 - x_1x_3x_4x_5x_6 +$$
$$+ 2x_1x_2x_3x_4x_5x_6. \tag{6.161}$$

Taking the mean of expression (6.161), we get

$$F_S = P(S = 1) = E_S = 2F^6 - 5F^5 + 2F^4 + 2F^3 \tag{6.162}$$

which is equivalent to the reliability expression derived in Section 6.2.

The failure probabilities of the components are generally small. Consequently, the failure probabilities of the cut sets are also small and the probabilities associated with cut sets intersections are even smaller. Therefore, an approximate direct method of computing system unreliability is to sum up the failure probabilities of the cut sets, neglecting the terms associated with the cut sets intersections from the inclusion-exclusion expression. We may then write :

$$F_S = P(S = 1) \lesssim \sum_{i=1}^{m} P(S_i = 1) = \sum_{i=1}^{m} F_{i_1}F_{i_2} \ldots F_{i_{n_i}}. \tag{6.163}$$

Expression (6.163) gives an upper limit of system unreliability. If the failure probabilities of the cut sets are smaller than q and the cut sets are mutually disjoint, the approximation error is less than $q^2C_m^2$.

This approximate method may not be applied in the case of the logical model, although the structure function is similar to (6.159), because the probabilities of the path sets are close to unity, and the probabilities of path set intersections may not be neglected.

We shall apply the approximate method to the communication system with structure function (6.160). For equal failure probabilities of the components we get, according to relation (6.163)

$$F_S = P(S = 1) \lesssim \sum_{i=1}^{4} P(S_i = 1) = 2F^3(1 + F). \tag{6.164}$$

An upper limit of the error is :

$$\sum_{j<k} P[(S_i = 1)(S_K = 1)] = 5F^5 + F^3. \tag{6.165}$$

For $F < 0.1$, the error is less than $5 \cdot 10^{-5}$ which is, however, greater than the value predicted in [45] because the cut sets have common elements.

We next consider the fault tree analysis of renewal systems. In the first step, only component renewal is considered and the system is viewed as non-repairable, with preventative maintenance. In this step, the system failure probability must be derived from the component unavailabilities. In the second step, system renewal is also considered and the analysis should yield a relation between system and component unavailabilities.

As we have seen in Section 6.3, an exact expression for the unreliability of systems with preventive maintenance may be derived only on the basis of the Markov chain model.

In the second step, we are tempted to use the structure function (6.159) to derive the relation between system unavailability and component unavailabilities. Indeed, when the renewal possibilities at the system and the component level are considered, we would then have

$$\bar{A}_S = P(S = 1); \quad \bar{A}_j = P(x_j = 1) \tag{6.166}$$

and we might apply a general method to derive the relationship

$$\bar{A}_S = \psi(\bar{A}_1, \bar{A}_2, \ldots \bar{A}_n). \tag{6.167}$$

If we apply the algebraic method and use logical-algebraic equivalences, the logical function (6.159) is transformed into the algebraic function

$$S = \sum_{i=1}^{m^*} C_i x_{i_1} x_{i_2} \ldots x_{i_{n_i}} . \tag{6.168}$$

Each term of expression (6.168) is a cut set, not necessarily minimal. The number of terms m^* in expression (6.168) differs from the number m of minimal cut sets.

Taking the mean of expression (6.168) we get

$$\bar{A}_S = \sum_{i=1}^{m^*} C_i \bar{A}_{i_1} \bar{A}_{i_2} \ldots \bar{A}_{i_{n_i}} . \tag{6.169}$$

Expression (6.169) is identical with the unreliability expression when no maintenance is considered.

However, expression (6.169) is not correct, because the states at instant t of the components in a cut set are not mutually independent. When a minimal cut set included in a non-minimal cut set of expression (6.169) causes system failure to occur, the other components in the non-minimal cut set remain dormant until the system repair is completed, so that they do not evoluate independently of the components in the minimal cut set.

Let us consider, for instance, the fifth term ($- x_1x_2x_3x_4x_5$) of expression (6.161). This is a non minimal cut set, since it contains the minimal cut sets $x_1x_2x_5$ and $x_1x_2x_3x_4$. Assume that the minimal cut set $x_1x_2x_5$ causes system failure. Then, components 3 and 4 cease their renewal processes until system repair is completed, so that the variables x_3 and x_4 are dependent on the variables x_1, x_2, x_5.

Relation (6.169) is exact only in the case of parallel systems with a number of servers at least equal to the number of components.

An exact analysis of maintained systems based on the fault tree model is not possible, and we have to invoke the Markov chain model in order to perform it.

Fortunately, an approximate fault tree analysis of maintained systems is straightforward [45]. Let $\bar{A}^{(i)}$ be the unavailability of cut set i. Since the cut set is equivalent to a parallel system, we may write :

$$\bar{A}^{(i)} = \bar{A}_{i_1} \bar{A}_{i_2} \ldots \bar{A}_{i_{j_i}} . \tag{6.170}$$

The result (6.170) is exact when there are enough servers, and approximate otherwise. If the cut set unavailabilities are small, the system unavailability may be approximated by their sum, since it is the probability of a reunion of events :

$$\bar{A}_s \simeq \sum_{i=1}^{m} \bar{A}^{(i)} . \tag{6.171}$$

Let $\{\lambda_{i_k};\ k = 1, 2 \ldots j_i)$ be the failure rates of the components in the cut set i. We shall derive the type I renewal density of a cut set. Failure of cut set i occurs if $j_i - 1$ components of the cut set have already failed at t and the remaining component, which is good at t, fails in $(t, t + \Delta t)$. We may then write :

$$h_1^{(i)}(t) = \sum_{k=1}^{j_i} \lambda_{i_k} \bar{A}_{i_1} \bar{A}_{i_2} \ldots \bar{A}_{i_{k-1}} A_{i_k} \bar{A}_{i_{k+1}} \ldots \bar{A}_{i_{j_i}} =$$

$$= \sum_{k=1}^{j_i} \lambda_{i_k} \frac{\bar{A}^{(i)}}{A_{i_k} \bar{A}_{i_k}} \approx \bar{A}^{(i)} \sum_{k=1}^{j_i} \frac{\lambda_{i_k}}{\bar{A}_{i_k}}. \tag{6.172}$$

The reliability function of a cut set may now be approximated by

$$R^{(i)}(t) = \exp\left[-\int_0^t \Lambda^{(i)}(u)\, du\right] \tag{6.173}$$

where $\Lambda^{(i)}(t) = \dfrac{h_1^{(i)}(t)}{A^{(i)}(t)}$.

The system renewal density is the sum of the renewal densities of the cut sets :

$$h_{1_S}(t) = \sum_{i=1}^{m} h_1^{(i)}(t). \tag{6.174}$$

The maintained system reliability function is the product of the cut sets reliabilities :

$$R_s(t) = \prod_{i=1}^{m} R^{(i)}(t). \tag{6.175}$$

We shall illustrate the approximate method by the example of the communication system with structure function (6.160). The system is maintained, and has identical components with failure rates λ and repair rates ρ. The minimal cut sets are :

$$S_1 = x_1 x_2 x_5;\quad S_2 = x_1 x_4 x_6;\quad S_3 = x_1 x_2 x_3 x_4;$$
$$S_4 = x_1 x_3 x_5 x_6. \tag{6.176}$$

The unavailability of a component is, according to expression (6.140) :

$$\bar{A}(t) = \frac{\lambda}{\lambda + \rho}\left(1 - e^{-(\lambda+\rho)t}\right). \tag{6.177}$$

For large t, we may write :

$$\bar{A}(t) \approx \bar{A} = \frac{\lambda}{\lambda + \rho} = \frac{\lambda/\rho}{1 + \lambda/\rho} \approx \frac{\lambda}{\rho}. \tag{6.178}$$

The cut sets unavailabilities are:

$$\begin{aligned} \bar{A}^{(1)} &= \bar{A}^{(2)} \approx \bar{A}^3 \approx \frac{\lambda^3}{\rho^3} \\ \bar{A}^{(3)} &= \bar{A}^{(4)} \approx \bar{A}^4 \approx \frac{\lambda^4}{\rho^4}. \end{aligned} \tag{6.179}$$

The system unavailability is then:

$$\bar{A}_S \approx 2\bar{A}^3 + 2\bar{A}^4 = \frac{2\lambda^3}{\rho^3} + \frac{2\lambda^4}{\rho^4}. \tag{6.180}$$

The type I densities are, according to relation (6.172):

$$\begin{aligned} h_1^{(1)} &= h_1^{(2)} \approx 3\bar{A}^2 \approx \frac{3\lambda^3}{\rho^2} \\ h_1^{(3)} &= h_1^{(4)} \approx 4\lambda\bar{A}^3 \approx \frac{4\lambda^4}{\rho^3}. \end{aligned} \tag{6.181}$$

The availability of a cut set is close to unity, so that the reliability functions of the cut sets in time interval $(t, t + x)$ may be approximated, for large t, by:

$$\begin{aligned} R^{(1)}(x) &= R^{(2)}(x) \approx \exp\left[-\frac{3\lambda^3 x}{\rho^2}\right] \approx 1 - \frac{3\lambda^3 x}{\rho^2} \\ R^{(3)}(x) &= R^{(4)}(x) \approx \exp\left[-\frac{4\lambda^4 x}{\rho^3}\right] \approx 1 - \frac{4\lambda^4 x}{\rho^3}. \end{aligned} \tag{6.182}$$

The system renewal density is:

$$h_{1_S} \approx \frac{6\lambda^3}{\rho^2} + \frac{8\lambda^4}{\rho^3}. \tag{6.183}$$

The system reliability function in time interval $(t, t + x)$ is, for large t:

$$R_s(x) \approx \exp\left[-\left(\frac{6\lambda^3}{\rho^2} + \frac{8\lambda^4}{\rho^3}\right)x\right] \approx 1 - \frac{6\lambda^3 x}{\rho^2} - \frac{8\lambda^4 x}{\rho^3}. \tag{6.184}$$

Relations (6.182) and (6.184) show a linear decrease of the reliability functions. This linear feature is restricted to small time intervals.

To complete the reliability analysis, we shall compute the statistical importance of the components. We shall use another definition of the statistical importance, more convenient than that applied in the logical model. The statistical importance of component k is the probability that component k had caused the system failure, given that the system has failed. If the definition refers to the first system failure, the component statistical importance is related to the system reliability, while if the system failure is not necessarily the first one, the statistical importance is related to the system availability.

Let α_k be the reunion of cut sets which contain component k. Component k causes system failure if a minimal cut set to which it belongs fails. According to the definition of conditional probabilities, we have:

$$I_{k_R} = \frac{\sum\limits_{k \subset \alpha_k} F^{(i)}}{\sum\limits_{i=1}^{m} F^{(i)}}; \quad I_{k_A} = \frac{\sum\limits_{k \subset \alpha_k} \bar{A}^{(i)}}{\sum\limits_{i=1}^{m} \bar{A}^{(i)}}. \tag{6.185}$$

Let us consider, as an example, component 3 of the communication system. Since it belongs to cut sets S_3 and S_4, its statistical importance is:

$$I_{3_R} = \frac{F^{(3)} + F^{(4)}}{F_S}; \quad I_{3_A} = \frac{\bar{A}^{(3)} + \bar{A}^{(4)}}{\bar{A}_S}. \tag{6.186}$$

If the system is not maintained, the statistical importance is related only to the system reliability. According to relations (6.164) and (6.186), we have:

$$I_{3_R} = \frac{2F^4}{2F^3 + 2F^4} = \frac{F}{1 + F}. \tag{6.187}$$

If the system is maintained we get, according to relations (6.179), (6.182) and (6.186):

$$\begin{aligned} I_{3_R} &= \frac{8\lambda\bar{A}^3}{6\lambda\bar{A}^2 + 8\lambda\bar{A}^3} = \frac{4\bar{A}}{3 + 4\bar{A}} = \frac{4\lambda}{3\rho + 4\lambda} \\ I_{3_A} &= \frac{2\bar{A}^4}{2\bar{A}^3 + 2\bar{A}^4} = \frac{\bar{A}}{1 + \bar{A}} = \frac{\lambda}{\lambda + \rho}. \end{aligned} \tag{6.188}$$

6.5 Concluding remarks on structural analysis

The necessity of a systematic presentation led us to define the models described in this Chapter on the basis of the system structure function. Intuitive models, such as graphs and combinational circuits, were associated with the structure function. Thus the quantitative analysis was simplified, as the minimal path sets or cut sets were given from the beginning. Actually, the structural model is not derived from abstract functional relationships, but by a quasi-empirical analysis of the real system, which shows up unfavourable combinations of primary events. Such a procedure yields the intuitive model directly and the structure function is derived from it. There is not yet a general algorithm for building up a structural model, though efforts have been made in this direction [44, 46].

Derivation of the structure function from the intuitive model follows the classical procedure of boolean functions minimization [47]. Once the minimal path sets and cut sets have been derived, the quantitative methods of analysis described in this Chapter may be applied.

The numerical computation of system reliability measures relies on the assumption that the reliability measures of the components are known. For a realistic appreciation of the component reliability measures, one must take into account the fact that these measures depend also on the role the component plays and the place it occupies in the structure of the system.

The system structure affects component reliability in two ways. First, by the stress to which the component is subjected. Second, by the failure criteria imposed by the system performance and structure. To account for these two effects, one must adapt the component reliability measures to the real stresses, according to the methods described in Chapter 4. Moreover, the component reliability should be estimated with respect to the real tolerance intervals which the system structure imposes on the component parameters. In the usual reliability analyses, this requirement is often ignored, so that component reliability data based on intrinsic failure criteria are used, inappropriately. Extrapolation of the reliability measures for different failure criteria is a difficult operation, which may be performed only under certain restrictive assumptions [48, 49]. When extrapolation is not possible, special reliability tests should be performed that take into account the actual failure criteria of the components.

The quantitative reliability analysis provides important results on the properties of coherent non-maintained systems.

Apart from the series system, all coherent systems with memoryless components are aging systems, since failure of some components brings the system into a wearing-out (worn) state, intermediate between the initial and the failed state. More specifically, systems with CFR components are IFRA. The IFRA property is also conserved under the formation of coherent systems, that is, coherent systems with IFRA components are also IFRA [50].

The aging property of a coherent system implies that its reliability function may fall below the reliability function of a CFR system, which is an exponential function. Consequently, preventive maintenance is required in order to hold the system reliability function above some specified level.

References

1. ZADEH, L.A., DESOER, C.A., Linear System Theory, McGraw-Hill, New York, 1963.
2. ZADEH, L.A., POLAK, E., System Theory (in Romanian), Ed. tehnică, Bucharest, 1972.
3. CĂLIN, S., BELEA, C., Complex Automatic Systems (in Romanian), Ed. tehnică, Bucharest, 1973.
4. DRUJININ, G.V., Reliability Theory of Radio-Electronic Systems (in Russian), Energhia, Moscow, 1976.
5. GEHER, K., Theory of Network Tolerances, Akadémiai Kiadó, Budapest, 1971.
6. RYABININ, I., Reliability of engineering systems. Principles and analysis, Mir, Moscow, 1976.
7. DRUJININ, G.V., System Reliability (in Romanian), Ed. tehnică, Bucharest, 1969.
8. MIHALACHE, A., A system theory approach to reliability modelling, Proceedings Relcomex, Wroclaw, 1980, pp 25—39.
9. GROSH, D.L., A parallel system of CFR units is IFR, IEEE Trans. Reliab., R-31, 4(1982) 403.
10. BILLIKAM, J.E., Comment on "A parallel system of CFR units is IFR", IEEE Trans. Reliab,, R-33, 5(1984) 410.
11. TERZIAN, R.C., Comment on "A parallel system of CFR units is IFR", IEEE Trans. Reliab., R-32, 1(1983) 36.
12. KLEFSJÖ, B., On closure under formation of parallel systems, IEEE Trans. Reliab., R-33, 5(1984) 423.
13. FAURE, R., LAURIÈRE, J. L., Fiabilité et Renouvellement des Equipements, Gauthier-Villars, Paris, 1974.
14. SINGH, C., A cut set method for reliability evaluation of systems having s-dependent components, IEEE Trans. Reliab., R-29, 5(1980) 372—375.
15. CĂTUNEANU, V.M., MIHALACHE, A., IONESCU, I., Reliability Analysis of Systems with Certain Types of Dependency between Component Failures. Researches in Electronic Technology and Reliability (in Romanian), Ed. didactică și pedagogică, Bucharest, 1979, pp. 164—175.
16. BANERJEE, S.K., RAJAMANI, K., Closed form solutions for Delta-Star and Star-Delta conversions of reliability networks, IEEE Trans. Reliab., R-25, 2(1976) 336—339.
17. SINGH, C., KANKAM, M.D., Comments on "Closed form solutions for Delta-Star and Star-Delta conversions of reliability networks", IEEE Trans. Reliab., R-25, 3(1975) 110—111.
18. GUPTA, H., SHARMA, J., A Delta-Star transformation approach for reliability evaluation, IEEE Trans. Reliab., R-27, 3(1978) 212—214.
19. GADANI, J.P., System effectiveness evaluation using star and delta transformations, IEEE Trans. Reliab., R-30, 1(1981) 43—47.
20. LOCKS, M.O., Recent developments in computing of system reliability, IEEE Trans. Reliab., R-34, 5(1985) 425—437.
21. ABRAHAM, J.A., An improved method for network reliability, IEEE Trans. Reliab., R-28, 1(1979) 58—61.
22. LOCKS, M.O., Recursive disjoint products, inclusion-exclusion and min-cut approximations, IEEE Trans. Reliab., R-29, 5(1980) 368—371.
23. BENETTS, R.G., On the analysis of fault tress, IEEE Trans. Reliab., R-24, 3(1975) 175—185.
24. RAI, S., AGGARWAL, K.K., An efficient method for reliability evaluation of a general network, IEEE Trans. Reliab., R-27, 3(1978) 206—211.
25. AGGARWAL, K.K., RAI, S., Symbolic reliability evaluation using logical signal relations, IEEE Trans. Reliab., R-27, 3(1978) 202—205.
26. TIWARI, R.K., VERMA, M., An algebraic technique for reliability evaluation IEEE Trans. Reliab., R-29, 4(1980) 311—313.
27. SCHNEEWEISS, W.G., Calculating the probability of boolean expression. IEEE Trans. Reliab., R-26, 1(1977) 16—22.
28. SATYANARAYANA, A., PRABHAKAR, A., New topological formula and rapid algorithm for reliability analysis of complex networks, IEEE Trans. Reliab., R-27, 2 (1978) 82—100.
29. SATYANARAYANA, A., A unified formula for analysis of some network reliability problems, IEEE Trans. Reliab., R-31, 1(1982) 23—32.
30. HWANG, C.L., TILLMANN, E.A., LEE, M.H., System reliability evaluation. Techniques for complex/large systems — a review, IEEE Trans. Reliab., R-30, 5(1981) 416—423.

31. AHMED, S.H., A simple technique for computing network reliability, IEEE Trans. Reliab., R-31, 1(1982) 41—44.
32. SATYANARAYANA, A., HAGSTROM, J.N., A new algorithm for the reliability analysis of multi-terminal networks, IEEE Trans. Reliab., R-30, 4(1981) 325—334.
33. IMAGAKI, T., HENLEY, E.J., Probabilistic evaluation of prime implicants and top-events for non-coherent systems, IEEE Trans. Reliab., R-29, 5(1980) 361—367.
34. MIHALACHE A., VOLOVICI, D., A Dynamic Method for System Reliability Analysis (in Romanian), Proceedings CNETAC, Bucharest, 1984, pp. 215—217.
35. SOBOL, I.M., The Monte Carlo method, Mir, Moscow, 1975.
36. KUMAMOTO, H., et al., System reliability analysis and diagnosis by Monte Carlo method, Proceedings IFAC, Helsinki, pp. 1613—1620.
37. EATON, M.C., LANG, C.K., Sequential destruction method for Monte Carlo evaluation of system reliability, IEEE Trans. Reliab., R-29, 1(1980) 27—32.
38. EVANS, R.A., Independent or mutually exclusive, IEEE Trans. Reliab., R-29, 4(1980) 289.
39. MIHALACHE, A., CORLĂŢEANU, V., Application of the General Structural Models to the Reliability Analysis of Repairable Systems (in Romanian), Buletinul IPB, XLV (1983) 87—95.
40. KUMAMOTO, H. et al., State-transition Monte Carlo for evaluating large, repairable systems, IEEE Trans. Reliab., R-29, 5(1980) 376—380.
41. NAKAGAWA, T., OSAKI, S., Stochastic behaviour of two dissimilar-unit standby redundant system with repair maintenance, Microelectron. Reliab., 13(1974) 143—148.
42. NAKAGAWA, T., OSAKI, S., Optimum preventive maintenance policies for a 2-unit redundant system., IEEE Trans. Reliab., R-23, 2(1974) 86—90.
43. CHU, T.L., APOSTOLAKIS, G., Methods for probabilistic analysis of noncoherent fault trees, IEEE Trans. Reliab., R-29, 5(1980) 354—360.
44. LEE, W.S. et al., Fault tree analysis, methods and applications — a review, IEEE Trans. Reliab., R-34, 3(1985) 194—203.
45. FUSSEL, J.B., How to hand-calculate system reliability and safety characteristics, IEEE Trans. Reliab., R-24, 3(1975) 169—174.
46. LAPP, S.A., POWERS, G.L., Computer-aided synthesis of fault tress, IEEE Trans. Reliab., R-26, 1(1977) 2—13.
47. NAKASHIMA, K., HATTORI, Y., An efficient bottom-up algorithm for enumerating minimal cut sets of fault tress, IEEE Trans. Reliab., R-28, 5(1979) 353—357.
48. KOVALENKO, I.N., Complex Systems Reliability Analysis (in Russian), Naukava Dumka, Kiev, 1975.
49. CĂTUNEANU, V.M., MIHALACHE, A., Extrapolation Measures for Different Failure Criteria. (in Romanian) Researches in Electronic Technology and Reliability. Ed. didactică şi pedagogică, Bucharest, 1981, pp. 266—274.
50. BARLOW, R..E., PROSCHAN, F., Statistical Theory of Reliability and Life Testing. Probability Models, Holt, Rinehart and Winston, New York, 1975.

Chapter 7

Principles of reliability

We shall end our presentation of reliability theory and practice by an attempt to integrate the ideas developed in this specific technical field into the general context of scientific thought.

Reliability theory has been developed from a quantitative point of view to provide a measure of the capacity of a system to resist natural degradation and a prediction of system behaviour. Consequently, reliability control in the cybernetic sense has been implemented by evaluating the technical effort required to bring reliability to an optimal level.

Mathematical modelling gave a rigid meaning to the concept of reliability, reducing it to its quantitative measure. No model perfectly fits the reality it seeks to describe, and it is therefore always necessary to distinguish between the phenomenon and the model, between the physical object and its measure.

The terminology itself fosters confusion between reliability as a general property of a system to resist degradation and its mathematical model, which uses the reliability function as one of the main reliability measures.

The development of reliability models is based on the concept of failure, defined as the event which occurs when a system performance transgresses its tolerance limits. This conventional definition of failure assumes perfect knowledge of the relevant system performances and of the tolerance intervals associated with them. It is difficult to specify either which performances are relevant for a certain application or the tolerance intervals which would guarantee satisfactory system behaviour.

The ordinary meaning of failure is qualitative; it implies that the system becomes unfit to accomplish a specified operation. Failure may occur either by the slow drift of some relevant performance or by its sudden variation. In the latter case, there is a discontinuity between the failed state and the working state of the system.

In reliability theory, the notion of failure differs from its ordinary meaning by the rigour of its definition, which lends it a somewhat artificial flavour. Moreover, no theoretical distinction is made between "sudden" and "drift" failures, although sudden failures may have more dramatic consequences because the system may behave unpredictably.

The artificiality of the reliability model is common to all mathematical models, because they are based on a quantitative definition of a qualitative concept. Here, the qualitative meaning of failure as "unfit for use" is replaced by the quantitative meaning of failure as "tolerance limits transgression". This approach is essential in eliminating any ambiguity from the natural concept but it also severely restricts the significance of the operational concept [1]. The unambiguous definition of failure by tolerance limits transgression is much more restrictive than the natural concept of failure, defined loosely as unfitness for use. Thus we have to admit that the mathematical models developed throughout this book are valid only to the extent that the unambiguous definition of failure is sufficiently close to its natural meaning.

The reliability models are stochastical, so that prediction of the behaviour of a system based on its past evolution and/or its structure, is possible only in probabilistic terms. Reliability theory employs the probability in its three present meanings : classical, statistical and subjective.

The classical interpretation of probability is used when the fraction of failed items in a reliability test is considered. The true reliability function is then the fraction of items which survive from a collection subjected to the same stresses during a specified time span. Since we can seldom consider a certain collection exhaustively, the classical interpretation is not often used.

The statistical meaning of probability is the limit of the frequency with which an event occurs. This interpretation is valid in all reliability models when sufficient experimental data are available for predictions.

The subjective meaning of probability is considered unorthodox in many an empirical field [2]. However, it is useful in reliability studies, since quantification of the degree of belief is essential in using all available information when experimental data are scarce. The subjective interpretation generates a class of adaptive models which are continuously updated by Bayesian inference under the impact of additional experimental data. Bayesian inference is also used, but in a different manner, when the objective interpretation of probability is considered. Thus, the reliability case studies provide pertinent illustration of the probability interpretations and contribute to the clarification of the Bayesian controversy in science and engineering.

A natural question arises concerning the stochastical feature of the reliability models. The degradation phenomena may be essentially stochastic or may be governed by deterministic laws yet unknown, so that the stochastic feature of the models reflects only our ignorance concerning the profound nature of the degradation processes. A similar problem arises in quantum mechanics, where the search for "hidden variables" is intended to discover the determinism underlying the present stochastical models [3].

The failure mechanisms are based on physical and chemical processes which can be regressed down to the quantum level but this does not imply that the quantum indeterminism should be preserved at the macroscopic level which concerns reliability theory. On the contrary, assuming the validity of the stochastical models at the microscopic level, the degradation phenomena should appear as deterministic at the macroscopic level, by the action of the law of large numbers.

For this reason it is interesting to find conditions in which the degradation phenomena are governed by deterministic laws. Such laws should provide failure prediction only by system performance measurement.

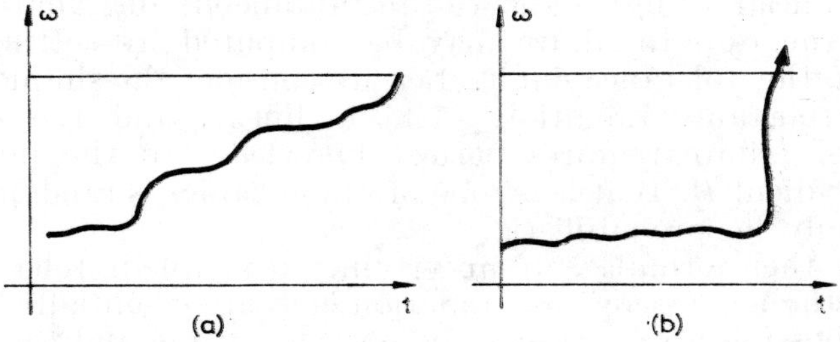

Fig. 7.1. — Failure process;
(a) continuous; (b) discontinuous.

A failure is defined as the transgression of the tolerance limits for any performance of the system. This event is more predictable if the performance changes continuously with time.

Figure 7.1 depicts the relevant performance ω of a system as a continuous or a discrete process. The process is not reversible, so that the continuous failure process is monotonic, on average.

More often than not, the two types of failure processes — continuous and discrete — are correlated with the different stages of system evolution. It is accepted that during the useful life, when the failure rate is constant, failure occurs by a sudden jump of system performance, while in the wearing-out period, failure is caused by a continuous drift of the performance. System failure is more predictable in the latter case, the failure process is considered completely random in the absence of aging and only relatively random for systems with "memory". Such considerations are erroneous because there is no correlation between the type of aging and the evolution of system performance. It is perfectly possible to produce an exponential model for the time to failure under the assumption of a continuous variation of the system performance, and IFR or DFR models are also compatible with discrete jumps of the performance. Moreover, failure remains a random event as long as the performance degradation is modelled by a stochastical process, irrespective of whether this process is continuous or discrete. However, it is true that failure is more predictable if the performance varies continuously since, in this case, more information on system reliability is available. A prior condition of a deterministic prediction is then the continuity of the relevant performances. In practice, if a performance is obviously discontinuous, it is nearly always possible to consider in its place a continuous variable sensitive to the failure process and strongly correlated with the system performance.

Under the continuity assumption, a deterministic prediction implies that the failure mechanism could be modelled by an equation of the type :

$$\omega(t) = A + \int_0^t B(t)\,\mathrm{d}t. \tag{7.1}$$

If both the initial value A and the instantaneous degradation rate $B(t)$ are known, the time-to-failure may be computed by setting expression (7.1) equal to the tolerance limit. Let us consider the simplest case when $B(t) = B =$ constant. Equation (7.1) is linear, and the deterministic prediction of failure requires perfect knowledge of the initial value A and of the gradient B. If at least one of these values is random, the prediction could only be probabilistic.

We can then formulate an uncertainty principle in reliability theory, stating that the macroscopic degradation laws are essentially probabilistic if it is not possible to know exactly both the value and the degradation rate of a system performance at any time instant t. This is analogous to the Heisenberg uncertainty principle in quantum mechanics, since the performance value is similar to the position and the degradation rate is similar to the speed of a particle.

The uncertainty principle for reliability theory does not seem so implacable as the Heisenberg principle. We believe that probabilistic and deterministic approaches may alternate in the natural evolution of reliability theory. The present probabilistic models could be replaced by partially deterministic models as more information on the failure mechanisms become available. The future deterministic models would then give way in their turn to more sophisticated probabilistic models able to describe new degradation phenomena, currently ignored. An infinite progress of knowledge by alternate steps, probabilistic and deterministic, could be the route for any science.

The reliability models should themselves be reliable and their validity must be tested, which is not straightforward, especially in the probabilistic case.

According to Karl Popper [2], a theory cannot be verified because, irrespective of a great number of facts which corroborate the statements of the theory, there is no absolute certainty on its validity and future refutation is always possible. A reliable theory should be falsifiable, that is it should be possible to oppose the statements of the theory with statements based on empirical observations which could possibly contradict the theory. A theory which cannot be contradicted because it adjusts itself to any new facts is unfalsifiable and, hence, unscientific. The models are reliable when they are able to stand repeated testing against empirical evidence.

It is obvious that the probabilistic models of reliability are not, strictly speaking, falsifiable in the sense of Popper's conceptions [2].

We shall examine the reality of these models more closely from the point of view of the requirements that any scientifical theory should meet.

Let us consider a typical statement of the reliability theory that is the interval reliability function of a system in $(t, t + x)$ is equal to R_0. This is a probabilistic prediction at time instant t of the future system behaviour, stating that an operation of duration x starting at t will, with probability R_0, be successfully accomplished. This statement may be interpreted in the following manner, using the classical meaning of probability. Let N be the totality of identical systems which start working at time instant t, having the same age at t and being subjected to identical stresses during the operation time interval. It is expected that $(1 - R_0)N$ systems will fail during the time interval $(t, t + x)$. The prediction may be rigorously verified by a direct experiment on the N systems. Actually, the difficulty of an experiment performed on an entire class of systems prevents validation of the prediction in practice.

If one uses the statistical meaning of probability, n operations of duration x are assigned to the system such that, after each completed operation, the system is turned back to its state at time instant t. If n is large, it is expected that about R_0n operations will be successfully accomplished. It is obvious that, however large n, we can never ascertain that the system reliability is precisely R_0. If m successes are observed, the frequency of success m/n is an estimate of the reliability function and there is no point in comparing it with R_0 in order to validate the prediction.

According to the estimation theory, the predicted value R_0 should be compared to the limits of a confidence interval which contains the true value of the reliability function, with probability $1 - \gamma$. If R_0 is outside the confidence interval, the prediction is considered false, although there is a small probability γ that it might not be so. Naturally, when the number of trials n is small, the confidence interval is large, and the validation of the prediction is meaningless. One may conclude that a probabilistic statement is more readily falsifiable if the accuracy of the estimation is high, that is, sufficient experimental data are available. This would be a truism if the estimation procedures (Chapter 2) were not capable of computing the experimental effort required for significantly validating a theoretical prediction. The probabilistic statements are reasonably falsifiable only when enough data are available, and the amount of data may be computed from the confidence level and the estimation accuracy measures. The confidence level provides a criterion to invalidate a statement with a specified probability that the true statement is erroneously considered false. The estimation accuracy specifies the amount of data required to maintain the confidence interval sufficiently small. The higher the estimation accuracy, the more readily falsifiable is the statement. One may compute, from the accuracy measure, the probability of validating a false statement, provided that the false statement is not a simple negation of the true one.

To be specific, let us consider true the statement "the system reliability function is R_0". Then, the statement "the system reliability function is not R_0" is false. It is not possible to compute from the estimation accuracy measures, the probability of considering true the false statement defined by negation, because it would imply taking into account the infinity of values different from R_0 of the reliability function. The probability of considering a false statement to be true may be computed only when the false statement is formulated as "the system reliability function is R_1 (different from R_0)", which is a rather restrictive case.

Since one may specify beforehand only the confidence level, that is, the probability of invalidating a true statement, it follows that it is more significant to prove the falsity of a statement than to prove its validity. However, the falsity of a probabilistic statement may be proved only if the amount of empirical evidence is sufficiently large.

This reasoning agrees well with Popper's conceptions which state that a theory is validated only by repeated attempts to prove its falsity. To each probabilistic statement, such as "The system reliability function in $(t, t + x)$ is R_0", another probability is attached, being the confidence level which, for a given confidence interval, is a measure of the validity of the statement. The statement is more readily falsifiable when the confidence interval is narrower.

The above considerations are based on classical estimation theory which uses the probability in an objective manner. The confidence level is the limit of the frequency for which a statement is considered true given that it actually is true.

Testing the validity of reliability statements based on bayesian inference follows a different line of reasoning. The statement "the system reliability is equal to R_0" is too rough in the bayesian context. It is preferable to express one's degree of belief on the value of the reliability function by attaching to it a subjective probability density function which could be maximum, say, at R_0. The confidence level is now a subjective probability which expresses the analyst's reasonable degree of belief in the statement "the system reliability function lies somewhere between R_1 and R_2". Such a statement cannot be falsified if the subjective pdf covers the whole range $(0, 1)$ of the reliability function. As more data are gathered, the initial statement adapts itself to the empirical evidence and a final subjective pdf results which is concentrated around the true value of the reliability function. A statement continuously formulated so that it would not exclude any improbable outcome is never false, as it tends to be true when experimental information is accumulated. The trouble is that a decision must be taken before sufficient information is available, so that the validity of a statement relies on a good initial guess and on a reasonable description of the a priori information.

Reliability theory is an attempt to ameliorate the natural degradation implied by the law of entropy increase. Let us consider a sample of similar systems. As they are not completely identical, their performances are random values initially distributed according to some probability density function. As the systems evolve independently of one another, the distribution of their performances at a latter time instant has a greater variance than the initial distribution. The increase of the performance variance is the way the entropy law manifests itself, with the effect of a decrease of reliability. An irreversible evolution to failure seems the way of all systems, and reliability theory should only try to predict when what is bound to happen will happen. However, the systems we are concerned with, be they technical, living or social, are open systems, so that their evolution to an improved state is nevertheless possible. A system could use the energy exchange with its surroundings to maintain its working state, or it might use adaptively the fluctuations of its variables in order to improve its internal order [4]. We have already seen that the open feature of technological systems may be used for upgrading system reliability by preventative renewal. Incorporation of artificial intelligence might make it possible for the systems to use the external perturbations in order to upgrade their performances, so that evolution towards failure would no longer be inexorable.

References

1. QUINE, W.V.O., Success and limits of mathematization, XVI Weltkongress für philosophie, Düsseldorf, 1978.
2. POPPER, K.R., The Logic of Research (in Romanian), Ed. ştiinţifică şi enciclopedică, Bucharest, 1981.
3. de BROGLIE, L., Certainty and Uncertainty in Science (in Romanian), Ed. politică, Bucharest, 1981.
4. PRIGOGINE, I., STENGERS, I., The New Alliance (in Romanian), Ed. politică, Bucharest, 1984.

Index